SPACEPORT **EARTH**

SPACEPORT **EARTH**

THE REINVENTION OF SPACEFLIGHT

JOE PAPPALARDO

Overlook Duckworth
New York • London

This edition first published in hardcover in the United States in 2017 by
The Overlook Press, Peter Mayer Publishers, Inc.

NEW YORK
141 Wooster Street
New York, NY 10012
www.overlookpress.com
For bulk and special sales, please contact sales@overlookny.com,
or write us at the above address

LONDON
30 Calvin Street
London E1 6NW
info@duckworth-publishers.co.uk
www.ducknet.co.uk

Cataloging-in-Publication Data is available from the Library of Congress

A catalog record for this book is available from the British Library

Book design and typeformatting by Bernard Schleifer
Manufactured in the United States of America
ISBN 978-1-4683-1278-2
ISBN UK: 978-0-7156-4586-4

FIRST EDITION
1 3 5 7 9 8 6 4 2

Man must rise above the Earth—to the top of the atmosphere and beyond—for only thus will he fully understand the world in which he lives.

—SOCRATES, philosopher

If there is a small rocket on top of a big one, and if the big one is jettisoned and the small one is ignited, then their speeds are added.

—HERMANN JULIUS OBERTH,
founding father of rocketry

A scientist describes what is. An engineer creates what never was.

—THEODORE VON KÁRMÁN,
physicist and aerospace engineer

One day I would love to do a rock gig on the moon. How rad would that be?

—TOMMY LEE, drummer, Mötley Crüe

CONTENTS

Map of US Spaceports

Commercial, Government, Private, and Proposed Launch Sites

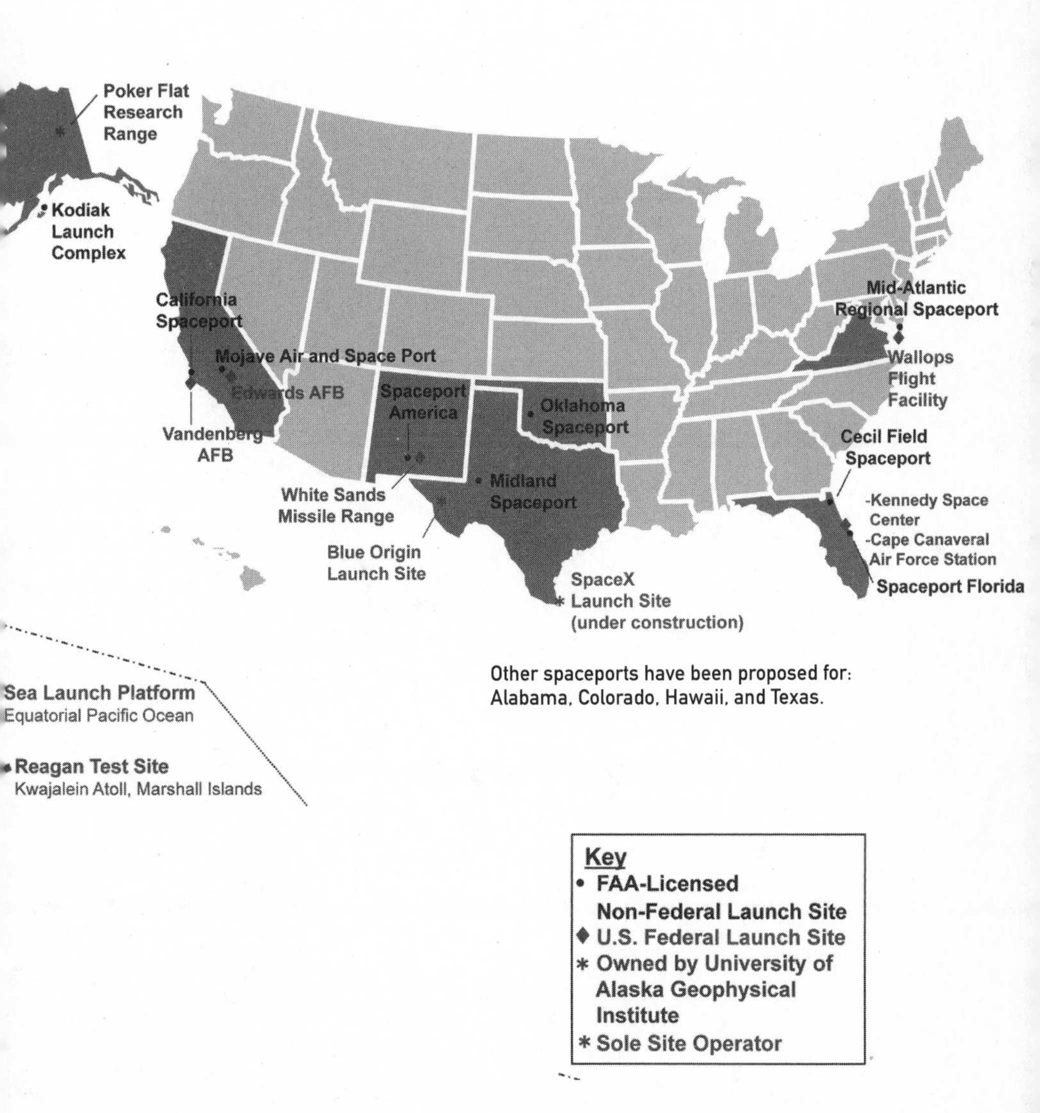

Illustration credit: Federal Aviation Administration

CHAPTER 1

ENDINGS AND BEGINNINGS

AFTER FIVE MINUTES IN THE ORLANDO AIRPORT, I WANT TO RUN for the exit and speed to the spaceport at Cape Canaveral. It is July 5, 2011, and I'm in Florida to witness the last flight of a US Space Shuttle. I imagine every moment at Kennedy Space Center over these next few days will be historic. The banality of the airport quickly becomes intolerable.

I'm here to see the end of American human spaceflight, or at least what everyone says is the end. NASA's attitude seems to promote the pervasive, glum feeling of an era ending. They bill the last launch of the Space Shuttle *Atlantis* as "The Grand Finale." This spacecraft has been flying my entire adult life, and now it is becoming obsolete. If I want to see this piece of aviation history anywhere but on display at a museum, it's now or never.

As an editor at *Popular Mechanics* lucky enough to have supportive bosses, I can often self-assign stories that I want to cover. So I credential myself through NASA's media relations team, call Hearst Corporation's travel office, start researching the hardware of a shuttle launch, and, almost as an afterthought, begin to explore the history of the spaceport here at Cape Canaveral.

Space Transportation System (STS)-135 will be the last shuttle launch, ever. But for me it will also be an inauguration. This will be the first space launch I'll ever see, despite the fact that I've been covering spaceflight as a journalist for more than five years, including stints at *Smithsonian Air & Space* magazine and my current gig at *Popular Mechanics*. I have walked the empty halls of failed satellite launch companies in Dallas, interviewed legends like Gene Kranz and Buzz Aldrin, and toured the laboratory workshops of the Jet Propulsion Lab's deep space projects. It wasn't the science but the engineering that first hooked me on the space beat: the dangerous vehicles, daring people who make and ride them, robotic expeditions to unfathomable environments, and high-stakes industries that formed around it all.

But for all I wrote about space, I have never seen an actual rocket launch, that first violent step needed to get up there, until now. To be honest, before today I thought about landing on other planets more than the brute effort it took to escape from this one.

Like a lot of people, the reality of the Space Shuttle's retirement snuck up on me. The spacecraft seemed to be with me my whole life. I watched launches live on TV and as a kid once walked across the street to the Suffern High School lawn to watch a dot of light go past. My father assured me that it was the shuttle zipping past in orbit. I snuck out of class to watch *Challenger* rise on live TV, and returned to my classroom in shock after it detonated in midair. I was amazed by the in-space fix of the Hubble Space Telescope, astronauts doing work while perched on the tip of robotic arms, the Earth rotating beneath them. From an apartment in New York City, I numbly followed the cable news coverage of the tragedy of *Columbia* cracking up during reentry and scattering herself across half of Texas.

My interest drifted as the shuttle made repetitive missions to the International Space Station—a modern marvel, to be sure, but essentially an achingly slow construction project best appreciated when it was completed. But the drama of retirement has reignited my curiosity, and I leave in the airport in Orlando and head due east with high expectations.

I learn something very quickly about Cape Canaveral while driving into Cocoa Beach: There's more going on here than just shuttle launches. Tall lightning rods mark the locations of other launch pads up and down the coast. The Air Force personnel at Patrick Air Force Base call this "heavy-lift country" because of the frequency of military and national security satellite launches that roar away from here every few months.

These routine aeronautical miracles don't get much attention in 2011—the launches are in the hands of a capable but costly monopoly called the United Launch Alliance (formed by aerospace giants Lockheed Martin and Boeing) and not worthy of much public discussion. That, however, is destined to change. My visit will cross the historic intersection of two epochs of spaceflight, but it's hard to think about that right now.

I pull the rental car into Cape Canaveral under ominously cloudy skies. Weather is the bane of all launches, no matter where the spaceport is located. The parking lot is a grass field located close to the Vertical Assembly Building (VAB). It's impossible to miss the VAB—a five-hundred-foot-tall box emblazoned with the NASA logo. The VAB is the largest single-story building in the world and has housed every manned space launch system since 1968. After being used to assemble Saturn rockets for the Apollo program, the massive structure became where engineers mated the Space Shuttle orbiters to their solid rocket boosters and external fuel tanks.

It stands on the horizon, visible for miles, forming a large white tombstone for NASA's manned flight program. With the shuttle program's end, the VAB will now be empty, purposeless, good only for the tour groups.

America has no replacement for the Space Shuttle and is about to lose its ability to deliver people to orbit. This fact is galling to me—how could NASA, and Congress, and a line of presidents drop the ball so completely? The smug chin scratchers who savaged the lunar Constellation program—including Aldrin, who lent his fame to support Barack Obama's push for cancellation—are now staring at *Atlantis* on the launch pad and saying, what's next? Aldrin wanted to trade the moon for Mars, but NASA seems to have let him down. How can we ever hope to get people to Mars when we can't even reach the International Space Station? The Constellation program was flawed, and wildly expensive, but at least it existed and had a destination.

Plenty of people blame the shuttle itself for the US's flaccid manned spaceflight program. They have good reason—it never reached its declared potential, and its prohibitive cost and tally of dead crew did immense damage to NASA's reputation. But here in Florida on the verge of its last flight, I don't want to hear any of that. This is an emotional pilgrimage for many, and I find that I'm not immune.

When I get to the Cape, the *Atlantis* orbiter is already at Launch Pad 39A. You can see it from miles away without binoculars, gleaming white despite the overcast skies. I want to see that powerful vehicle tear its way out of Earth's gravity well, with seven intrepid souls pressed into their seats, along for a heavy G ride. I want to see history. I'll get that—and an unexpected glimpse of the future as well.

Looking at the shuttle and the spaceport infrastructure that it supports, if someone were to tell me that private space companies would change the face of heavy-lift country over the next five years, I would call them naïve. But by the time I leave Florida, just a few days later, I am thinking about it differently.

IN MAY 1947, THE PENTAGON ATTEMPTED TO LAUNCH A V-2 ballistic rocket, like the ones the Nazis used from French soil to bombard London in World War II, outside White Sands, New Mexico. The weapon rose from the pad in a roar of flame and smoke, and leapt into the sky. Within moments, however, the testers knew something was wrong. The missile was heading south instead of north.

A V-2 is a pretty simple rocket: a tank with an alcohol-water mixture, another tank with liquid oxygen, a small chamber with hydrogen peroxide, and some pipes to mix it, ignite it, and route it through the engine. They are not known for their sophisticated guidance systems. (Just a couple of vanes on the fins.) The range of the V-2 can stretch hundreds of miles, and the Air Force was about to learn that the room they allotted to test them needed to increase.

The errant missile flew over El Paso, Texas and continued south, ultimately running out of fuel over Mexico. It careened into a cemetery in Juarez, leaving a deep, fifty-foot-wide crater. The crash in a populated area indirectly cemented the future of Cape Canaveral as a spaceport. The Pentagon fast-tracked a project to create missile ranges that fired experimental weapons over water. They wanted one on the east coast, and another on the west.

Every missile shot has what is called a footprint, a teardrop-shaped area under the flight path that needs to be clear of people in case the thing crashes. For rockets shot into space, that expanse usually is only found on coastlines, such as eastern Florida. In 1960, the Department of Defense noted in a report that the Atlantic Missile Range on Cape Canaveral "is substantially saturated with missile launching facilities and test instrumentation." The Air Force had a test range for its wonder weapons, and didn't want to share it.

That was bad news for a new rocket team that wanted to move into the Cape: the National Aeronautics and Space Administration (NASA). President Eisenhower founded NASA in July 1958 when he signed the National Aeronautics and Space Act in reaction to the Soviet launch of *Sputnik*, but he quickly tired of spending money on it. By 1960, NASA budgets had been slashed.

John F. Kennedy's election changed that course. His call to go to the moon was a calculated one. "He was not a visionary enraptured with the romantic image of the last American frontier in space," NASA chief historian Steven Dick puts it in his *50 Years of NASA History*. "Kennedy as president had little direct interest in the US space program."[1]

But Kennedy had a competitive streak and wanted to surpass the Russians, who beat America in manned spaceflight by lofting cosmonaut Yuri Gagarin into orbit. So Kennedy opened up the US Treasury

1. To find a real space supporter of that era, look to Lyndon B. Johnson, who in 1958 said, "Control of space is control of the world."

checkbook and spent lavishly on a manned trip to the moon, for purely political reasons.

From where would these massive rockets launch? Studies were commissioned and ideas put forward on places the government could build a new, civilian spaceport. Ideas rose and were shot down: construct an offshore site, a floating spaceport. Nah, too hard to maintain. Create one near Brownsville, Texas. Nope, too many populated areas on both sides of the border. How about a South Pacific island, or somewhere in the Bahamas? Too expensive to develop out in the middle of nowhere. The coast of Georgia? Not with that Intracoastal Waterway right there; rockets sailing overhead would certainly hinder boat traffic.

That pretty much left the Atlantic Missile Range. The military tenants hooted and hollered for a couple years, first trying to keep the civilian program away, then trying to make NASA a tenant of their facilities. In the end, officials gave the go-ahead for NASA to expand into its own operation on the Cape's Merritt Island. On November 28, 1963, six days after President Kennedy's assassination, President Johnson announced that Cape Canaveral would be renamed after the fallen president.

With that, the Kennedy Space Center became the US government's primary spaceport for the next five decades.

LAUNCH IS FRIDAY; ON WEDNESDAY VIPS AND MEMBERS OF THE media are scheduled to see *Atlantis* up close. We gather around the launch pad to watch as an incredibly complex gantry called the Revolving Support Structure (RSS) ease away from the launch pad. They call it a rollback, and it is the first major event that a member of the press covering a shuttle launch must attend. Everything is a ritual, following a checklist tucked in a blue NASA binder somewhere.

The overcast skies open up and rain pelts those waiting for buses to the launch pad. Those holding cameras tuck them under plastic ponchos and dash for nearby tents. The rain drizzles away, leaving everyone steaming hot, chafed, and frustrated. The clouds open again just before military security officials bring in an explosive-sniffing dog to check our gear.

Our destination is Space Launch Complex 39A, the most famous launch pad in history. Apollo 11 launched from here on the way to the moon on July 16, 1969. NASA repurposed it to handle the Space Shuttle, but a lot of the same hardware and spaceport operation ethos remain. When something works, the tendency is to do it the same way, over and over again. Which is fine, up to a point. The Apollo program was wasteful, disorganized, done with clumsy urgency—but it was staffed with brilliant people. The procedures they created remain legacies at the Cape for decades: assemble the rocket upright in the VAB and use a massive vehicle called a crawler to make a painfully slow roll to the pad. In some ways, Apollo bound all who followed at the Cape to its precedents, infrastructure, and methodology.

Any griping about security and weather ends when the bus lurches into motion. Dozens of poncho and rain-jacket clad journalists—six school buses full of them—become children on a summer camp field trip. The passengers are in the grips of nostalgia and getting charged up by their proximity to aviation history.

A good part of the road to pad 39A follows the loose-stone track taken by the crawler-transporter, the 2,700-ton behemoth that hauls the shuttle from the Vertical Assembly Building to the launch pad. We pass the forlorn, eight-tracked machine as we drive in, its last trip finished.

The bus pulls to a halt and we stampede out, with a NASA media minder shouting at us: "Remember, return to bus four. *Bus number four*!" We have less than a half an hour to spend next to the shuttle, and no one wants to waste any time inside the bus. The shutterbugs are the most jumpy, eager to make every second count.

Launch Pad 39A sits on the crest of a gradual but steady incline. A fence corrals us. A banner from NASA ground staff, decorated with stars and stripes, stretched across it reads: GO ATLANTIS!

The shuttle stands, stark white, atop the 390-by-325-foot concrete pad foundation. Such slabs are called hardstands, and they are high-tech, steel reinforced miracles of engineering. The shuttle's fuel tanks—one rust-orange main tank and two slender, off-white boosters—tower higher than the orbiter. This serves as a stark reminder that most of any launch vehicle's weight is in fuel and oxidizer storage. It takes a lot of energy to escape earth's gravity, and that means power, and power means lots of fuel.

The boosters are filled with an explosive powder. The main tank—the big orange one—holds pressurized liquid hydrogen kept at -253°C, and liquid oxygen at -183°C. These combine and burn in the orbiter's main engines. The main tank is empty, for now. It's too expensive to fill them early because the cryogenically cooled liquid oxygen boils off.

The spacecraft and tanks are massive, but more massive still is the RSS gantry itself. It's a solid, utilitarian structure, with two thick bridges connecting the gantry to the spacecraft. Umbilical lines run from building to launch vehicle. A lightning tower, meant to attract bolts that could strike the spacecraft, crowns the entire tableau, extending hundreds more feet into the cloudy sky.

Inside the RSS are clean rooms, where techs in white biohazard-like suits would work on payloads without contaminating them. Those rooms are empty now, never to be used again. I wonder what it would feel like to be the guy who turned those lights out for the last time.

It's hard to imagine the hulking RSS structure moving. Yet imagine it we must, since the actual rollback happened before we got there. The RSS had opened like a cabinet door that morning, disconnecting the workrooms and access points used to prep the shuttle.

No one seems to care that we missed the big reveal. Simply standing scant feet from a 180-foot-tall spacecraft is putting us over the edge. Video crews step into each other's shots. Everyone mugs for the cameras, like tourists. I start shaking hands like I'm running for office. I'm sharing this once-in-a-lifetime experience with strangers.

Thunderclouds build and roil overhead, but the trip to the pad gives me confidence. *Atlantis* is real now and is going to fly, and I am going to see it from roughly three miles away. The idea of watching the craft blaze into the sky becomes a trembling reality. How can something so *big* move so *fast*?

The half hour standing at Launch Pad 39A goes past quickly, too quickly. The media minders herd us into the buses with much yelling and waving of arms. The journalists file away begrudgingly, lingering by the bus doors until forced inside. I crane my head for one last look. I am no longer just an observer. I now have a vested emotional interest in seeing this spacecraft take flight. I miss her as soon as we start driving away.

Many that day doubt that Launch Pad 39A will ever be the spot of a significant launch again, but they are wrong. Eventually, NASA will pay $1.3 million to LVI Environmental Services of New York to remove the historic RSS infrastructure piece-by-piece, disassembling it slowly to avoid damaging the hardstand. They will be clearing the pad for a new tenant.

That tenant will be a scrappy space company straight out of a Robert Heinlein novel: Space Explorations Technologies Corporation. Or, as everyone calls it, SpaceX.

THERE'S A LULL AT THE KENNEDY SPACE CENTER MEDIA ROOM after the rollback finishes. The NASA control room staff ticks off items from checklists, base weathermen examine lightning detector data, and Air Force meteorologists launch the first of many balloons to measure conditions in the higher reaches of the atmosphere.

The press corps has little to do, however, so a field trip to the SpaceX launch pad fills the void in the schedule.

In 2011, when the shuttle program is winding down, SpaceX is still basking in the glow of a successful launch and recovery of its Dragon capsule, intended to one day deliver cargo to the International Space Station (ISS). SpaceX is one of two companies to have won NASA contracts to send unmanned cargo capsules to the International Space Station. (The other firm is Orbital Sciences Corporation in Dulles, Virginia.) This is a radical departure from NASA's business as usual. Traditionally, the agency would hire companies whose engineers would build rockets to NASA's specifications, with the understanding that Uncle Sam would be the only user of the design. It's a business practice followed worldwide.

These private space companies are doing something different, though. They design their own rockets, loading them with their own spacecraft, and running missions out of their own operations centers. They work for the government but own the blueprints and intend to rent rides into space to anyone else with the cash to pay. By doing so, the iconoclasts who own private space companies are reinventing spaceflight.

SpaceX is already pulling ahead of other competitors, and the company's owner, PayPal billionaire Elon Musk, is making wild boasts about tenfold reductions in launch costs with no loss of reliability. I hope for the best, but deep down have always taken him for a blowhard. But by 2011 he's come farther than I thought possible, and these recent successes give me pause.

Our tour starts at SpaceX's Florida offices, just outside the spaceport's gates. The gumdrop-shaped Dragon capsule, SpaceX's first successful spacecraft, sits on display under a tent, proudly showing the dark scars of its reentry. Spacecraft coming back to earth travel so fast that the air molecules rushing around it rub, heat up, and form superhot plasma. Plasma is gas that acts like a liquid, and as the altitude drops a teardrop of glowing, 3000°F gas surrounds the spacecraft. Visible char from this process is a badge of honor; it means this vehicle has been to space and back. The reporters, stacked red carpet–style behind a nylon strap partition, furiously snap pictures.

The bus shows up to ferry us into the security zone, to SpaceX's facilities inside the spaceport. Access to spaceports is always limited, protected by barbed wire fences, intrusion sensors, and security teams on land and in the air. Rockets run on dangerous and volatile things, often stored in pressurized containers, and the possibility of industrial and international espionage adds another layer of paranoia. These trappings only make it cooler for the journos on the bus.

Just before we pull out of the parking lot, a woman in a polo shirt and sunglasses hops on board: Lori Garver, Deputy Administrator of NASA and a key supporter of the agency's experiment with private space companies. She starts chatting with SpaceX Vice President of Communications Bobby Block and some beat reporters as the bus heaves forward.

Garver is the Obama administration's point person on spaceflight. Without the shuttle heading to the ISS, private space cowboys like Musk are about to become the faces of American spaceflight. For the time being, Garver is their ranch manager and tenacious defender. Many in Congress are not thrilled about the new public-private partnership, some made squeamish by fiscal responsibility, others protecting their job-creating constituents. Inside NASA, change equals risk. Many in NASA see private space as a smokescreen behind which Washington could retreat from American-manned spaceflight. Nearly the entire retired astronaut core rebelled publicly when the initiative was announced.

We roll to the SpaceX launch pad, visible at a distance even though there is no rocket on it. Four tall lightning towers stand at each corner, grounded by metal cables. A massive sphere filled with 110,000 tons of liquid oxygen sits like a monstrous golf ball; SpaceX bought the tank from an Air Force scrapyard literally for one dollar. This liquid oxygen will be mixed with rocket-grade kerosene to power SpaceX's rocket engines.

This is Space Launch Complex 40. Of course, everyone calls these SLCs, pronounced "slicks." SLC 40 has witnessed launches since 1965, sending Titan rockets off the planet by the dozens. It fell silent in 2005, until SpaceX arrived looking for a pad to launch its Falcon 9 rocket, part of its NASA contract to supply the ISS. The unmanned capsule outside SpaceX's local headquarters blasted off from this spot.

This is no longer a NASA launch pad, and the infrastructure here shows that this company is doing things differently than any other launch provider. They have abandoned the idea of vertical assembly used by launchers worldwide. SpaceX assembles its rockets horizontally, so there are no massive launch towers or vertical assembly buildings here like the ones that mark the skyline in other parts of Cape Canaveral.

There are no crawlers to ferry upright rockets at a single mile per hour to this pad, either. Instead, rockets travel on a short rail line to the transporter-erector vehicle. The machine slides out from a preparation facility, holds the rocket horizontally, and then maneuvers it upright on the launch pad. The short-term benefit of the system is that if a launch is delayed, SpaceX can easily lay the rocket down and slide it back into the hangar to fix the problem. In such a system, the same pad could (one day) fire rockets into space in quick succession. Musk designs for the future, not just the day's demands.

The tour guide for the rocket assembly area is Scott Henderson, director of Mission Assurance and Integration at SpaceX. A Falcon 9 rocket lays on its side inside the building adjacent to the launch pad. The "9" stands for the nine specially designed Merlin engines dropped in the rocket's first stage.[2] The Falcon moniker comes from the *Millennium Falcon* of *Star Wars*. Once upon a time rockets took names of gods: Apollo, Mercury, Atlas. Now they take on names of a new pantheon, drawing from a new cultural mythos.

Henderson points out the rocket's unique features, including those Merlins. It is the nation's first new large liquid-fuel rocket engine to fly in forty years. Russian engines are renowned for being better than American-made varieties, so most space firms to this point have bought foreign and slow-footed replacement development efforts. But SpaceX's engine is as innovative as its launch pad. The Merlin runs on highly refined kerosene and cryogenically cooled liquid oxygen. It uses a single injector, unlike more complicated engines

2. A quick word about rocket stages: A rocket will drop empty fuel tanks on its way into space, and the engines along with the tanks. Each section that jettisons is called a stage; most heavy launch rockets have three stages.

that mix fuel and oxidizers at multiple points. To an engineer, less complicated means fewer breakdowns and easier repairs.

A close look at the rocket's body reveals more innovation. The upper stage of SpaceX's Falcon 9 separates with a mechanical ball-and-socket system instead of using industry-standard explosives. Pyrotechnic charges have been used since before Apollo to decouple the empty parts of a rocket, since empty fuel tanks are just unwanted weight. But the explosives need to be defused if a launch is canceled, adding expense and delay. SpaceX's design negates those concerns, and enables the quicker turnaround of launch vehicles. This is what happened during the company's first, aborted launch to the ISS in 2010.

SpaceX doesn't just want to launch rockets for NASA, nor does Elon Musk dream of solely launching rockets for the US government. He plans to revive the commercial space launch market—lofting satellites for communications firms—that collapsed in the late 1990s. There hasn't been a commercial sat launch from a US company since the early 2000s, and SpaceX wants to reverse that trend, using the Falcon 9 and its heirs to do so. It's a wildly ambitious plan.

After the tour and return to the media room, it's time to share the experience. I send photos and captions of SpaceX's launch facilities to the home office to post on *Popular Mechanics*'s website. I get a call soon after from Katherine Nelson, a beleaguered but talented SpaceX spokesperson. "There's a problem with the stage separation hardware photo," she says. "You weren't supposed to take any pictures of that." I feel like an asshole, but I point out that Henderson never said I couldn't shoot it. Nelson doesn't care too much about that: "The problem here is that image violates ITAR."

Some acronyms carry some weight on the defense and national security beats, and ITAR is one of them. It stands for International Traffic in Arms Regulations, a set of federal rules that limit the sale and even exposure of foreigners to military equipment. In this case, I am showing the world a novel way of creating an intercontinental ballistic missile (ICBM).

Oops.

I hang up and send a bemused but urgent email to the web editor, and the photo disappears. To my knowledge, the North Koreans, Russians, and Iranians still use old-school explosive charges to separate rocket stages. Everyone else does, too. Old habits are hard to break.

MY CHEAP HOTEL IN ORLANDO SEEMS LIKE A FOUR-STAR RESORT ON Thursday night. To be honest, I need a break from the humidity, the press gaggle, and the sheer exhaustion of having to turn my mind into a recorder of fleeting historical events. Orlando is not very convenient for the assignment, but there are no rooms available close to the Cape.

I could have slept in my car at KSC to avoid the horrendous traffic jams associated with shuttle launches, but I'm expecting delays. Before I left the Cape, the weather officer gave a 70 percent chance that incoming storms would prevent launch. On my tour, I saw a bulletin confirming a lightning strike less than a mile from the launch pad. But by the time I'm in my room, NASA officials in charge of the launch are making confident noises about a break in the weather and an on-time launch. Their hope is to pop *Atlantis* off before the storms really get bad. So I settle into the hotel to write, keeping one ear on the radio.

There is only one sure way to know if NASA is serious about a launch the next day. If I hear that the pad technicians are filling the fuel tanks, it's a go. It takes hours to load the shuttle's 500,000 gallons of liquid oxygen and liquid hydrogen, so the choice to fill the tanks will have to be made before 2:00 a.m. to stay on schedule.

I snap the laptop closed and lay down at 1:00 a.m. My eyes are just slipping shut when the radio announcer cuts into the local news blather for a presser at Cape Canaveral. The NASA guy says something like, "We might be crazy, but we're going to go for it. Tanking will begin at 2:00 a.m."

My weary eyes flutter open. It's time to make that hour-long drive to the spaceport again. Rushing is necessary because of the legendary traffic jams associated with shuttle launches, and this one will be worse than any other. Not only is the press here in droves, but one million spectators are expected to watch the launch across the east coast.

With the crew scheduled to board the shuttle at around 5:00 a.m., it is conceivable that I'll be in an epic traffic jam staring at taillights instead of watching the final public appearance of these astronauts. I also have to secure desk space within the media center, which is teeming with a global sampling of radio, newspaper, wire service, television, and magazine journalists.

Even as I rouse myself to leave, I know there will be an eight-hour vigil spent in the Cape Canaveral bubble. A lot of this time will be spent outdoors, being bothered by mosquitoes, gnats, professional photogs, and aerospace history fanatics. And with each passing minute, I'll have to bear the increasing stress of something going wrong and cancelling the whole show.

I run through a supply list in my head as I pull on my pants. I need bug spray, food, and a gallon of drinking water. There are food trucks at the launch site, but the lines have been awful. On Thursday I had run into a college buddy, Chris Carroll of *Stars & Stripes*, covering the event, so I nab the twelve-pack of Budweiser from the hotel fridge that I plan on sharing with him.

At a gas station deli, I load all my provisions into a new Styrofoam cooler, next to the beer, and place the cooler in my rental car's trunk. There's already a steady stream of traffic on the road. The westbound lanes of State Road 528 are dark, but a glistening line of headlights stretches behind me in the rearview. A mini-migration of red-eyed and unwashed space geeks started at the same time I did, once NASA announced they would start filling that orange main tank.

The line of cars thickens as I turn onto route A1A. Some people are securing their viewing spots at public beaches, others stay with me as I headed into Kennedy Space Center. On Thursday afternoon the grass parking lot in front of the VAB had a dozen vehicles; today it looks like the grounds of a music festival.

I turn off the engine and close my eyes. They don't reopen for about an hour and a half.

When I wake up, the main tank is still being loaded. Filling that sucker takes just under three hours, finishing ahead of schedule at 4:48 a.m. By then, I've joined the press corps queuing up to board buses taking us to the next big milestone of a launch—the crew walkout.

The buses unload us with plenty of time for the photographers to set up ladders and get into position. "We've had people knocked off those ladders," one media handler warns us as we disembark. "Try to all work together, help each other out." Other well-wishers join the media, including NASA officials and the flight's medical crew. Nichelle Nichols, *Star Trek*'s Lt. Uhura, shows up looking lovely, posing for photos and demurely shaking hands. Bomb-sniffing dogs arrive, those mission-essential canines, and they snort over the media's belongings, the Airstream, and everything else that will get close to the shuttle crew.

The wait is not a long one. Within a half an hour, four NASA employees walk out carrying bowling ball–shaped spacesuit helmets. The astronauts appear minutes later to appropriate cheering, applause, and the whirr-snap of the cameras. This is the last public appearance before they enter the spacecraft. If all goes as planned, by this afternoon these four people will be in orbit, and soon floating inside a space station.

They pause, scheduled now to have an iconic moment. They wave and smile at everything and nothing. After a predetermined amount of time, they silently step into the Airstream—another throwback to the 1960s pedigree of the US space program—and roll off to SLC 39A.

The Airstream convoy includes an armored vehicle and a helicopter, which circles overhead. Security remains tight, as if the shuttle was to move through downtown Baghdad instead of a secure airbase in Florida.

THE ENSUING SIX-HOUR WAIT IS TAKEN UP BY EMPTYING BEER CANS, taking naps, and conducting occasional interviews of opportunity. I enjoy the reunion with Chris; we sit together and knock back a few cans of beer in my car. We rehash our favorite story from college, which peaks with me driving Chris's car south on I-35 in Texas with a torn-free gas pump nozzle and several feet of hose trailing from the car. Yeah, that was a good one.

Not too long after, I sit with Rear Admiral Craig Steidle (Ret.), the president of the Commercial Spaceflight Federation. CSF is the lobby and advocacy group for the private space movement. Without the shuttle, NASA has to rent rides to the ISS from Russia to launch people into space: $63 million per seat, at six seats a year, for five years. Steidle, like many, hates this fact and sees private space as a solution. He is a pioneer of the private space movement and did a lot of work as a NASA associate administrator in the Bush administration to begin privatization of NASA launches. A 2004 *New York Times* account of his appearance at an air show reveals how far he was from NASA's traditional model. "First order of the day for Mr. Steidle, a retired rear admiral, was a presentation in which he hung an 'open for business' sign on NASA," it reads.

In 2004 few in the government (at least publicly) were talking about flying astronauts on private spacecraft. Now it's 2011, and those nebulous plans are becoming reality. Design is becoming actual hardware. SpaceX's Dragon capsule is unmanned, but Musk made sure it was constructed with windows. There's no doubt he wants to fly astronauts—from NASA, private companies, or just himself. And he isn't alone.

It occurs to me that spaceports, old and new, are about to become very busy places.

Back in the media room, scientists jockey for attention. Space experiment demonstrations appear in the press room. One of them catches my eye: a box, just a little larger than a steamer trunk, called the Robotic Refueling Mission. The experiment on board *Atlantis* will prove that a robotic arm, teleoperated by humans on Earth, can service satellites in orbit. Such an arm must be able to open valves, disable safety features like wires and valve caps, and insert fuel lines into receptacles. When the module gets to orbit, astronauts will secure it outside the ISS, where a two-armed robot named Dextre will test it in space.

Outside, a NASA trailer opens to sell guests *Atlantis* memorabilia: coffee mugs, fridge magnets, books, T-shirts, mission patches, and jackets. I grab a t-shirt that says, "STS-135: The Grand Finale." I picture myself wearing it proudly, letting everyone know I was a witness to history, but like all such cheeseball memorabilia, I will never end up wearing it in public. (My wife will wear it to bed sometimes.)

All this is distraction. *Atlantis* stands on the horizon, and she is the only thing out there. I mean, you can see the gantry, fuel tank, and boosters, but when you look out over the three miles of swamp and lake at 39A, all you can really focus on is the bright white outline of a spaceplane glowing on the horizon.

At 6:45 a.m. the weather reports are improving. Launch director Mike Leinbach gives the launch team some encouraging words. He must know that hundreds of thousands of people are also eager to hear this over the public radio channel: "We have a shot at this one today."

The clouds break occasionally and bathe the launch pad with shafts of light, giving the early bird photographers something iconic to shoot. At no point in the last twenty-four hours have there been fewer than a dozen telephoto lenses trained at the shuttle. If a bolt of lightning strikes the shuttle, there is someone on hand to capture the event for posterity.

A common misconception about launch countdowns is that there is only one. In practice, they routinely start and stop at critical moments, called planned holds. We shuttle watchers endure each of them at T-11 hours, T-6 hours, T-3 hours, T-20 minutes and T-9 minutes. They don't usually resume until all the flight staff check in from various departments in a steady cadence of affirmations across the radio.

A high-pitched whine overhead draws attention. There goes four-time shuttle astronaut Rick Sturckow, streaking over the launch site in a T-38 jet to gauge the conditions. He lands, then hops into a Gulfstream-2, modified to fly like a landing Space Shuttle, and takes off again. If the shuttle develops a problem, one thing it could attempt to do is turn around and land at Kennedy. Sturckow's second flight determines that the shuttle can make it. If it couldn't land, and a distressed shuttle threatened populated areas, a mission flight control officer would press a self-destruct button. Spaceports retain the fearful respect of rockets born of the V-2 incident. They can kill them if need be.

By 10:00 a.m., about an hour and a half before launch, all the astronauts load into the shuttle, one by one. “Hatch is closed and latched for flight,” says closeout crew chief Travis Thompson, to a smattering of applause from the now densely packed throng on the grass.

The tension and the crowd grow, minute by minute. People are milling around, talkative, bragging of the number of launches they’d seen. I tell one of the photographers, wielding a camera telephoto that looks like a cartoon version of a musket, that this is my first.

“Better late than never,” he says.

At 11:00 a.m. things are still go for launch. I can’t believe it, refuse to believe it. I know a scrub could come at any minute.

Thanks to my early morning ride, I have a good view at the fence line. My binoculars are trained on the pad, and I try not to blink. I know I should look away once the engines ignite, or else I’d see nothing of the launch but an amorphous afterimage on my cornea.

At one minute, the arm atop the shuttle swings clear—the gaseous oxygen vent arm now unneeded.

“It’s gonna happen, baby,” the photog breathes, crouched over his camera.

I drop the binoculars to the ground and put my hands on my head like I’m holding my scalp in place. The crowd starts counting down the final thirty seconds. A swirling cloud of white vapor jets from beneath the craft. There’s an explosion of yellow, a bright, sunshine color that hasn’t been seen before on this drab morning. It cuts through the black and white canvas of my memory like a bright paint stroke. I swear I feel a warm flash on my face. The roar does more than hit us—it washes over us, shakes our guts, and fills our ears with the ripping sound of violently parted air.

The shuttle punches through a gap in the clouds, then vanishes. The column-like plume of its exhaust rises over the scene, a towering structure that slowly, stubbornly dissipates into the humid air. I stare at the plume as a reminder that what I just saw was real. No one in the crowd remains stoic. Some hug, others clasp hands, and a lot of men and women cry.

The minute the heroin wears off, a junkie wants another hit. After the launch I stop wondering about the crew of four who are in orbit, and I start wondering about what's next. I don't want to feel sad about the demise of the space program. More than anything, I want to see another launch. It's the start of an obsession and a purpose. STS-135's plume hasn't fully dissipated when I realize how much weight the private space movement needs to shoulder. I see the planet differently, as a cage that needs to be escaped by using all the skill, ingenuity, and explosive daring our species can muster.

That day, I start to see Earth as one big spaceport.

SPACEPORTS EXIST AT THE NEXUS OF MORE THAN EARTH AND SPACE. They exist in a fragile balance of risk and reward, industry and science, physics and politics. Everything is risky when it comes to rocketry—not just to life and limb, either. The upfront costs are measured in the hundreds of millions of dollars, the competition is extreme, and a launch failure is unforgivable.

What's not to like?

In the years since the STS-135 mission liftoff, I've used every connection and excuse to see rocket launches. I've seen blastoffs rise from jungle spaceports, watched rockets rise working US spaceports on both coasts, and witnessed ICBM launches from California into the Pacific. (Anyone who says a nuclear warhead isn't a spacecraft knows very little about them.) I've seen the Cape light up with old rockets of 1960s design and host new rockets and spacecraft that could usher in a golden age.

I'm not the only fool for spaceports. Communities all over the country are jockeying to create them. It's not just about jobs. Towns, counties, and states crave the prestige that comes from hosting such a wondrous place. Legends walk in spaceport towns. Strange figures with high IQs and jackets with mission patches wander through the local shopping malls and eat next to you at restaurants. Tourists come and stay at rocket-themed hotels. And every so often the citizens of a spaceport town get the best fireworks display on the planet.

Each budding spaceport, from Waco to Wallops Island, has its own story. Not just of its past, but a story of its potential future, resplendent with launching rockets, landing spaceplanes, grinning astronauts, and a tax base supported by moneyed nerds with advanced degrees.

The truth is that there is a global boom in spaceports, but most of them will never see an actual launch. The industry is already littered with drained back accounts, long runways cracked with weeds, office buildings gone silent, and rocket test stands rusting in deserts. Civic leaders are not discouraged, though, and some of them have learned some lessons from this boom/bust cycle by using the spaceport label to create nodes of aerospace science that could help provide tech jobs.

Still, spaceflight has never been more important to our lives, from monitoring nuclear missile launches worldwide to beaming Howard Stern into car radios. This wide array of uses is attracting big money—and big risks.

This is not the first time people have been predicting private space will take off, but this time promises more. More change, more opportunities, more companies and more of an impact on everyday Earthlings.

Frank DiBello, the CEO of the state-run agency Space Florida, says the current crop of players have an approach that's suited to the high-risk, high-cost world of spaceflight. "Before, it was hardware companies seeking to become information companies in space," he says. "Now, it's the opposite. IT companies are seeking to become space companies. Paul Allen, Jeff Bezos, and Richard Branson see a future in the next frontier. These people understand markets."

I get charged up being close to this limitless imagination and ambition. *Moby-Dick*'s narrator, Ishmael, goes to sea to reestablish his connection to the planet and himself. He enters a port town and his mood immediately alters, poised on the brink of an ocean that Melville calls a “harborless immensity.” Spaceports are even more dramatic gateways between realms and places where the cosmic immensity seems a step closer.

These are also the places where mankind can prove he’s clever and worthy enough for survival. We know that all life on Earth will die when our sun runs out of fuel, if not sooner. The only hope for our species is to spread beyond our planet, as professional spacers from St. Petersburg, Russia, to Mojave, California, are willing to remind anyone who’ll listen.

It can be hard to remember that optimistic ambition is at the foundation of all spaceports. You can get bogged down in discussions of budgets, economics, chemistry, national priorities, mathematics, tax incentives, and environmental reviews. But there are those who look up, past the atmosphere and into the deep black, and see a challenge worthy of the human spirit. They then create the world’s most powerful, dangerous vehicles to break out of our native planet’s gravity well, a high risk for a slight chance at building something else out there. And their dreams begin or end at a spaceport.

AFTER THE LAST SPACE SHUTTLE LAUNCH, I DECIDED TO PUT myself on the spaceport beat. Human spaceflight is only one aspect of the story. I quickly learned that there is an industrial battle for satellite launches underway, and that the status quo is about to be shaken.

To start my real education about how spaceports work and how the launch industry grew to its current form, I needed to embed with the best in the business. And that meant heading to the equator. The destination: the jungles of French Guiana, South America.

CHAPTER 2

WELCOME TO THE JUNGLE

I'M SITTING ON TOP OF AN ENORMOUS TRACKED VEHICLE CALLED the BV-206, churning along a muddy path through the jungles of French Guiana in 2012. This is no theme park, with carefully placed plants and plastic fronds, speakers hidden as rocks and animatronic monkeys. The real jungle has layers of dense foliage that devours sunlight, an ever-present soundtrack of insect noises, and a rich, musky smell that overwhelms the senses. And the monkeys in this canopy will throw things at you.

My escorts are a platoon of thickly built, heavily armed members of the French Foreign Legion. They hail from across the globe—this platoon has members from Ireland, England, Belarus, and China—but only French citizens can be officers.

The BV-206 is made to operate in mountains, and in that environment having the top of the engine jutting inside the cab is probably appreciated. But in the jungle, it makes the cab an oven. So I spend most of the patrol on top, surveying the scenery and listening to jungle survival tips from the Legionnaires.

"Everything is difficult in the forest," says Foulques de Samie, 1st lieutenant of the 1st platoon, 2nd company, before handing me off to subordinates. "Everything is dangerous. You have to know yourself very well to survive here."

Things turn from prehistoric to science fiction with the sight of a white cylinder standing more than 190 feet tall. "There's the baby," one legionnaire says as he glimpses it through a break in the trees.

The baby is an Ariane 5 rocket, sitting on a launch pad just outside the town of Kourou. The platoon's mission for the next few days is to protect this rocket and ensure no one molests the launch of the satellite in its tip, worth hundreds of millions of dollars. The soldiers do this by patrolling the jungle outside the spaceport in BV-206s and, if needed, by fast-roping from helicopters to intercept intruders.

More than half the world's satellites—communications, weather, scientific, and surveillance—have blasted off into orbit from this spaceport, located at the northeast coast of South America. About 47 percent of American-owned satellites have launched from French Guiana, but most people couldn't locate this spaceport or the country it's located in on a map.

Even the Legionnaires are essentially unimpressed with the job. "Protecting the launch area is not so interesting," says de Samie. "We do it because it's our job and our responsibility. But we prefer the other part of our mission, deeper in the jungle."

That "other part" is hunting illegal gold miners. The fight between *garimpeiros* and the Legion has freshly escalated. The week before I arrived, a group of miners armed with assault rifles ambushed French police, killing two.

"Normally *garimpeiros* just run or give up," says one legionnaire on the patrol, who declines to give a name. "So the problem is not necessarily their weapons, but that they want to kill us."

But the Legionnaires and general public can remain as ambivalent as they want; anyone in the space game realizes that this dramatically isolated spaceport is a vital foundation of the industry. The people who know about this place are space nerds, telecom company executives, billionaire investors, and satellite entrepreneurs. The people who rely on this place, this place they never heard of, tally in the tens of millions.

To understand how the spaceport works, think of it as an airport. The French government built this spaceport in the 1970s and the

European Space Agency runs it now. Astrium, a European firm now part of the EU aerospace giant Airbus, built the Ariane 5 rocket. The customers supply the payloads. The French-based multinational company Arianespace is the airline that buys and operates the rockets, books the tickets, and supplies customer support, including insurance and financing. There are no passengers; not a single person has ever reached space from here.

This is the industry standard commercial spaceport, and it has earned the title. When billionaire entrepreneurs and pundits talk about disrupting the market for commercial satellite launches, they are talking about taking on Arianespace here in South America. And these guys are very good at what they do, despite the fact that their spaceport is located in an entirely inaccessible jungle, thick with disease-spreading mosquitoes and renegade gold miners.

THE ROAD TO THE SPACEPORT IN KOUROU BEGAN IN ALGERIA. THE French established a rocketry program in 1947, centered at a missile and rocket range in the Sahara called Hammaguir.

Like other nations, the civilian and military rocket efforts grew into separate branches. Charles de Gaulle formed a national civilian space agency in 1962. But for France this schism coincided with another—the independence of Algeria. The new nation had no interest in keeping the Hammaguir open, and diplomats struck a deal to evacuate the facility in 1967.

Even as their sounding rockets rose from the desert, the French looked for a new spaceport. They had specific criteria for a new location that still serve as a pretty good checklist for any facility that aims to send things to orbit from earth:

- *Does the site have the potential for placing satellites on both polar and equatorial orbits?* Not all orbits are created equally. If the satellite passes over both poles, it can "see" almost every part of the Earth, which is rotating below. Every hour and a half, the satellite completes an orbit. Equatorial orbits circle the waist of the planet, so the satellite will scan the same area many times. If the rocket places its payload at a sweet spot 22,200 miles high, it will match the planet's spin and offer steady coverage of the terrain below.

This is a game of angles, so it matters where the spacecraft launch. The payloads are aimed at specific spots in space, measured in degrees from the equator. So the polar orbit has an inclination of 90 degrees. Those who are mathematically inclined will make a connection between the satellite's inclination (90 degrees) and the time it takes to circle the planet (90 minutes). The geostationary satellite will have an inclination of zero degrees, appearing as if it's in a fixed point in the sky.

A spaceport that can reach all of these orbits will have more customers, based on where they want their birds to go. Being close to the equator enables the widest possible selection of angles.

- *How close is it to the equator?* There's another reason to put your spaceport on the equator. The earth rotates at a little more than 1000 mph, so why not use that motion to help lift a rocket? Wonks call this the equatorial slingshot, and it makes a difference. The bump in speed means a rocket will need less fuel, about 15 percent less, to get its payload into space. The decrease in fuel means a smaller rocket is needed, and that saves even more fuel. Total fuel savings can reach 25 percent. The French knew all of this, making French Guiana's 5.3 degree latitude a nice option.

- *Is the area clear of people, buildings, and vehicle traffic?* Spaceports require elbow room. There is an amazing amount of power and violence in an orbital rocket launch, and when all goes well it should be harnessed and aimed in one direction. But when something goes wrong, it can be catastrophic for any surrounding community. The entire footprint—all the surface under the flight path of the rocket's first moments of flight—must be clear of people. The shape of this footprint looks like a teardrop, expanding its area the higher the rocket flies.

- *Is there a deepwater port with sufficient handling facilities or an airport with a landing strip capable of receiving long-range aircraft?* Logistics are vastly important to a spaceport, and that means infrastructure. The supply chains that feed spaceports often span the globe, so you need to be ready to accommodate industrial-size shipments of supplies and specialty cargo delivered in some big- ass vehicles. The French knew from the start that they'd be moving rocket parts and their customer's satellites from offshore.

- *Is the spaceport located somewhere with some political stability?* After Algeria, this makes particular sense to the French. Spaceports are not fly-by night enterprises, so some long-term thinking about their relationship with the locals is required. But it doesn't take a revolution to upend a spaceport. When it comes to stability, other spaceports are finding more mundane politics can interfere with the most ambitious plans. Local noise ordinances can derail a test schedule as easily as a technical mishap.

French Guiana fits every box of this checklist. Located on the east coast of South America, it has deepwater rivers that connect to coasts that enable shipping, a sparse population,[3] and enough unoccupied real estate to expand. Following the construction of a good-sized airport with a long runway, French Guiana can readily receive oversized cargo from customers.

French Guiana, however, has a bad reputation. People who know it tend to cringe when they hear the name spoken aloud. France's most infamous prison, the subject of the nonfiction book and film *Papillon*, existed on a trio of islands that sit off the coast near the spaceport called the Salvation Islands. Devil's Island, where officials sent inmates (often

3. There may be few of them, but in 2017 the local population staged a strike that shut down spaceport operations. Protestors claimed years of under-investment; calm was restored after France offered its territory an aid package worth billions of Euros.

political prisoners) to solitary confinement, is the most notorious of the three.

"When you say Guiana in Europe everyone thinks of *Papillon*, of the green hell," Pierre-François Benaiteau, director of the Guiana Space Center tells me. "No one is aware of *our* life here."

The Salvation Islands look close to shore, but it takes about an hour to get there by speedboat. The name dates to 1763, when France's first colonists arrived. They died in droves on the mainland, and the survivors fled the disease of the jungle for the islands, where they were eventually rescued, setting the tone for the colony's grim history. One reason the islands made such formidable prisons is the currents that sweep around them. Signs warn visitors of slippery rocks and deadly swimming conditions. Zhang Tong learned this the hard way. In 1997, the president of Great Wall Industry Corp., which builds and launches rockets for China, strayed too far to the water's edge and slipped in. His body was never recovered. A memorial stands near the place he disappeared; Arianespace officials say his family returns every year to place a wreath.

The first Ariane rocket launched in 1979. The rocket's creators designed it specifically to ferry satellites for paying customers. The spaceport has the same ethos, making the Guiana Space Center the first commercial spaceport in history. Nearby Isle Royal has a tracking station, built amid the old prison buildings, which the French space agency uses to monitor launches with radar. A network of these lonely tracking stations stretches across the globe, watching rockets over thousands of miles as they streak toward orbit.

Kourou is not on the way to anywhere except space. Getting a human being there—for example, a New York City–based journalist—is an exercise in patience. Jumping off from Miami isn't tough, except if the aforementioned journalist forgets to get a yellow fever vaccine. To get one requires a frantic taxi ride through Miami to make a crash appointment with a needle in a shady medical clinic.

The flight to Cayenne is a series of island hops, landing in beautiful harbors like Martinique and nightmarish airports like Port-au-Prince.

The stop in Haiti requires a re-boarding, complete with an inspection of all passengers' luggage, done by hand and in public.

Getting the rockets to the equator is not much easier. The rocket stages are too big to be flown in, so they have to be shipped. There's a shipping port located on a muddy river near the spaceport. Ships pull in, with their holds containing sections of the rockets. The ships open like a Higgins boat (i.e., a drawbridge-style door opening from the bow) and trucks roll in to haul rockets to the spaceport.

Some things have to be made on-site. The booster fuel is a fine powder of ammonia perchlorate, manufactured in French Guiana. No port or ship is eager to transit the powerful, explosive material in such vast quantities. So the French mix the materials together at an isolated plant in the jungle.

Satellites fly first class. They're carefully packed in climate-controlled containers and flown in whatever airplane can accommodate them. Really big communications satellites are usually transported by monstrous Anatov cargo airplanes. Motorcycle police and Arianespace cars make other drivers pull over as the crates are taken from the airport to the spaceport on the backs of trucks. Woe to any anacondas that slither onto the road when a convoy passes.

THE SINNAMARY RIVER WINDS MORE THAN THIRTY MILES THROUGH French Guiana, flowing north to the east coast of South America. Within six miles of the coast, the river widens before spilling into the Atlantic. That river mouth seemed inviting to the French settlers who scouted the river in 1624.

They were mistaken. Hostile reception from the Portuguese and indigenous peoples doomed the settlement, which was abandoned. In 1664 the French tried again, with more success, and founded the second colony in what would be French Guiana.

Colonization happens slowly here. Part of it is the unforgiving geography, but those who choose to build here have ambitions that take time and effort to achieve. It's hard enough to build a port to ship gold or rubber from the jungle; imagine the logistics needed to carve a spot of jungle into a spaceport.

Nearly four hundred years after the founding of a permanent town, a team of 350 Russian engineers starts clearing out a patch of rain forest near Sinnamary. The 2007 project began with an eighteenth century vibe. The land was surveyed, cleared, and connected with mud roads. The engineers plotted the route for a half-mile rail line to ferry massive equipment to the remote location. It took four years and hundreds of millions of dollars, but slowly a new launch pad rose from the jungle. In a triumph for stubbornness and diplomacy, the site witnessed its first Soyuz launch in 2011, just a few months before I visited the site.

The Russian launch pad near Sinnamary is six miles north of the site used for the Ariane 5. Having the Soyuz in French Guiana is a coup for Arianespace, which can now arrange—and get a cut of—those satellite launches. Arianespace guards against rocket difficulties with the ability to switch vehicles, including the Russian options and a smaller rocket developed by the Italian Space Agency. Launch pros call this "schedule assurance," and it's a big deal. Small payloads can hitch a ride on smaller rockets or tuck into larger rockets as secondary payloads.

A near-straight road extends from the Ariane launch site to the adjacent two buildings. Escorted and chauffeured by company staff, my photographer and I pull into the Russian facility, passing through both spaceport and Russian security stations along the way. There are giant coffins near the office buildings—sealed containers holding rocket stages, piled in stacks. The volume indicates the Russians are here to put things into space.

There's a noticeable difference between the French and Russian attitudes here at the spaceport. The Russians glower the whole time I visit. All of a sudden I remember all the negative comments from defense officials and pundits regarding the Galileo satellite network, the EU-Russian venture to reduce dependency on US-run GPS. Some Pentagon officials mused in public about the potential need to shoot one down during future conflicts.

The grim attitude is something the Europeans tell me not to take personally. The Arianespace engineers are too diplomatic to badmouth the Russians, but they do cop to "difficulties" and "cultural differences in the way we do things" getting in the way.

The Russians are stubborn partners. Case in point, the Soyuz launch pad. It's stoutly built, a slab of concrete hollowed with pipelines and electrical connections. It's designed like the ones Russia uses at its own spaceport in Baikonur, Kazakhstan. There is one massive pit to channel the flames and sound waves away from the rocket.

By contrast, the Ariane 5's pad uses underground tunnels (flame trenches) to rout the fiery exhaust and a cascade of water to dampen the acoustic waves. It's too cold for such a system on the steppes of Kazakhstan, so the Russians use massive concrete slopes. For the sake of consistency—and only consistency—they do it that way in French Guiana, too.

Still, the launch pad is a little different from those found in Kazakhstan. There's a massive building above the pit to protect the rocket from the high humidity. But the real difference is invisible: a gravitational slingshot that gives the Soyuz the ability to carry 3.3 tons of payload at the equator versus 1.8 tons for the same launch in Russia.

For me, the presence of Russians only enhances the mystery and appeal of French Guiana. The jungle spaceport belongs on the cover of a vintage sci-fi book. It has that special blend of high-tech and frontier grit that can only exist in remote places. The town of Kourou includes bars stocked with off-duty Legionnaires, Brazilian prostitutes, and transvestites. Anacondas slither across the highways, and sharks cruise the shoreline. Foreign Legion units shelter in forgotten buildings amid the toxic debris of sounding rockets. Restaurant walls are lathered with mission stickers. And every so often, the entire town gathers by the shoreline to picnic together and watch a rocket streak into space.

Pierre-François Benaiteau has been at the spaceport for more than twenty years, watching it grow from an isolated facility with a single launch vehicle to a cornerstone of the space industry with square footage as big as Manhattan. He can see two launch pads from his office window.

The rooftop of the office building is a popular place to watch the rocket move from its assembly building to a launch pad. They work on the rocket while it's upright, then use a massive crawler to slowly ferry it to the launch pad. It takes hours to get there. They've done this for decades. They see no reason to change.

Arianespace is the best in the world because they are so traditional, predictable, inside the box. They don't name their rockets after *Star Wars* vehicles. Innovation in spaceflight is always associated with greater risk, and that is a problem for those who pay tens of millions of dollars to deliver stuff safely into space. To a professional, none of this is supposed to be sexy.

"I don't like science fiction," Benaiteau says.

IT'S IMPOSSIBLE NOT TO BE IMPRESSED STANDING AT THE FOOT OF A 193-foot-tall rocket, especially on the day it will launch into space. Like the Space Shuttle, the Ariane 5 rocket uses twin boosters that lift the craft off the pad and fall away when empty. Fuel makes up 90 percent of the rocket's weight.

On the day before the launch, the pad is not as static as it appears. Deep booms rumble every couple of seconds as cryogenically cooled liquid hydrogen boils off. There is a steady hiss of helium running through the fuel lines to keep them pressurized.

We tour the pad with Clay Mowry, president of Arianespace's American subsidiary, and Charlie Ergen, a billionaire whose satellite is onboard. The two have known each other for years.

"That helium is a rare commodity, but this close to the launch it has to run all the time," Mowry says.

Ergen turns his head. "So is that why these launches cost so damn much?"

Joking at the launch pad releases some of the tension; this is high stakes science. The payload is one of the biggest communications satellites ever launched, worth $350 million. A failure during launch could mean many months of delay. Every second it stays on the ground, money it can be generating for the company is being lost. The meter is running.

The launch control center is built and sealed like a bomb shelter. The cars in the parking lot are always backed into spots in case a quick getaway is needed. On launch day, the closest Arianespace allows visitors to get to liftoff is an overlook called Toucan, situated on top of a hill about three miles away. It's so close that if a catastrophic failure happened, poisonous fumes from the rocket fuel could overcome guests. Lots of gas masks are available, with an obligatory tutorial on their use. The VIPs—mostly customers—often gather at Toucan to watch their cargo leave for orbit. Charlie and his wife Candy brought four of their children to this launch. The daughters, all college-aged, wear dresses for the occasion. The boyfriend of one daughter gets a crash course in orbital dynamics from Mowry, who proves to be an able teacher.

The sun is getting low on the horizon. The launch is less than fifteen minutes away.

Ergen is famous for being a professional poker player, and his face is calm considering he's about to watch hundreds of millions of his dollars violently blasted from the pad. Candy, a co-founder of the business and veteran of launches from every spaceport in the world (including China and from a rocket-launch ship at sea) remains poised as well.

She chats calmly with the assembled company reps and satellite people, whose shared business interest has morphed over the years into an extended, convivial community. "I get nervous at every launch, every one," she admits to me. "But seeing launches from every provider has been the highlight of my professional life."

Final countdown commences. With a plume of billowing white steam and smoke and a bright flash, the Ariane 5 is aloft. The boom of the boosters igniting comes seconds after we see it; the shock waves wash over the crowd. We hear the distinctive tearing sound of the rocket parting air, reaching the speed of sound in just thirty-four seconds. The white-yellow tail is too bright to watch under binoculars.

The rocket rises overhead and moves out to sea toward the islands, leaving a thick white trail behind that plays off the sun's rays. Under binoculars, the single point of light divides into three. Twin embers die out to either side of the pinprick of light; the boosters run out of fuel and fall away. Moments later, the rarest spectacle of a launch: a glimmer of white and the light dividing again. The fairing has opened, and it too is falling back to the sea.

"That's great, it's really rare to see that," Mowry says, smiling. He should grin—his rocket has worked as planned.

Charlie Ergen is now glued to the television screen showing the rocket's planned trajectory—a climb followed by an extended burn across the globe, picking up the full benefit of the planet's gravitational slingshot before it careens upwards. Sunset turns to night and bats careen under the awning, hunting bugs. Mission control confirms the main engine has run its course.

"MECO," Charlie says, glancing at his wife.

"Main engine cut-off," she responds, passing back his look. It's an oddly intimate moment.

At twenty-four minutes, mission control will know if the satellite was released successfully. As that countdown progresses, Charlie Ergen switches his weight from sandaled foot to sandaled foot. Word comes in: Separation is a success. "Good job boys," he says, then reins it in out of respect to the other payload. There's another team nearby waiting to hear if their weather satellite is safely away.

After it does, the real celebration begins: rum drinks and Cuban cigars at the hotel, enjoyed while watching the chain of navigation lights blinking along the route to Devil's Island. Despite the exotic setting, it's a low-key party given the monumental gamble of the day. But when you wager for truly high stakes, a quiet end of a successful day is the only reward needed.

IT'S HARD TO BE COMFORTABLE AT THE ST. PETERSBURG INTERnational Economic Forum in Russia. It's not just the heat or the language barriers. Coming to Vladimir Putin's soiree in 2014 carries more of a burden.

I had accepted the Russian government's invitation to moderate a panel at SPIEF, Russia's premier international business gathering, some years before. That trip produced some great information and contacts for the magazine. This time, the theme of the forum is "A World in Transition." It was an apt title, and the reason I don't say no. I want to see the growing US/Russia divide firsthand.

By April 2014, when SPIEF begins, relations between the US and Russia are rocky. The crisis in Ukraine prompts sanctions against Russia, NASA halts much of its cooperation with the country, and hundreds of executives of US companies sit out the forum. SpaceX founder Elon Musk and a NASA official are on the original roster of invited panelists for my session. Neither is here. I'm the only American on the panel—and likely the only one in the room.

My panel features A-list Roscosmos officials, an Indian space entrepreneur and two Europeans. The Europeans' attitudes interest me. It seems like tensions in Europe might make space cooperation a political casualty. EU-led regime change in Libya roils Putin, and his incursion into Syria brings Russia back to the front of Middle Eastern politics, but on the other side of the US/EU proxy fights there. The Russians also botch the launch of the first Galileo satellites, parking them into incorrect orbits from their spaceport in Kourou.

But the European panelists do not seem to consider a retreat. "Space activities and business and cooperation is above short-term political crisis," says Jean Loic Galle, president and CEO of Thales Alenia Space. "Even during what we called the Cold War, US and Russia continued to cooperate."

His firm partners with Russia on satellites and launch programs. Their most recent joint company formed in 2013. "Whatever the current situation, cooperation with Russia is very strategic for our company and we will try to expand it," he says. An executive with Airbus echoes his comments, saying space projects should "transcend politics."

If there's an exit button to the Euro-Russian space partnership, it won't be pressed for many years. Arianespace plans to operate Soyuz until at least the end of 2019. However, in the years to come the European companies will announce plans to create the Ariane 6, which could conceivably replace the Soyuz in 2020. The plans for new rockets, modern as they are, are still rooted in the old model of spaceflight—government-run space companies, allied with other nations to make money from commercial contracts.

In the United States something else is happening. The private sector is competing against these foreign companies and the governments behind them. While government-paid launches for NASA and the US Defense Department are lucrative, the growth area is lofting of communication satellites for well-heeled clients. It's a prospect that seemed outlandish no more than a decade ago.

But a handful of aerospace iconoclasts in the US are making noise, sounds that will be heard as far as Moscow and French Guiana.

CHAPTER 3

MOMENTS IN MOJAVE

Oak Creek Road winds its way outside the town of Mojave, California. A narrow ribbon heading east/west through the desert hills, the road leads straight into a line of wind turbines, dozens of them mounted in long lines.

Parked on the high ground along an access road, near the base of a steadily turning turbine, I can scan the town of Mojave. This windswept hamlet may only be a hundred miles north of Los Angeles, but it feels like an outpost on the edge of the world. There's not much here: 3,800 residents, more or less. The dusty main drag has two traffic lights, a cluster of fast-food franchises, and one decent roadhouse called Mike's, where miners, bikers, and pilots drink, shoot pool, and watch sports on ESPN.

On the northern edge of town, a chain-link fence marks the boundary of the Mojave Air and Space Port, which sprawls across 3,300 acres of desert. A control tower stands sentinel over three runways, the longest of which extends more than two miles. Weathered hangars, some dating back to World War II, line the main runway. A strange, bullet-shaped aircraft stands in a cul-de-sac by the airport's main entrance.

What goes on at this modest airfield makes Mojave the hub of global aerospace research and development. This place kickstarted an entire branch of the private space movement, even before it became the first commercial spaceport in the world and the first inland spaceport in the country.

Most hangar doors along the airfield are shut tight. Bizarre aircraft, secret Pentagon programs, and private spaceships take shape in these aluminum-sided buildings. The few that are cracked open offer glimpses of pressurized tanks, mechanics in oil-smeared overalls, and vehicles with smooth white fuselages, emblazoned with black tattoos mandated by the FAA: "Experimental."

This is 2008, three years before the Space Shuttle stopped flying and early in my tenure at *Popular Mechanics.* This is my first big assignment for the magazine, and Stu Witt unknowingly holds the future of my career in his hands.

Just a week before, I sat in the office of Jim Meigs, editor-in-chief of *Popular Mechanics*. He told me he had great language for a cover: "Inside the New Area 51." The only problem: He wasn't sure where that was.

So he asked me, his resident aviation geek. I knew of only one place: Mojave, California. Sure, it was a civilian-run facility and certainly not a secret. But the air base at Groom Lake, Nevada, was (or should have been) known as the cradle for groundbreaking aviation. In 2008, private space industry pioneers are the ones on the leading edge of aviation. And the epicenter of their most unique, experimental efforts is a small desert town in California.

The cover story has to be put together in days rather than the typical weeks. "How do we get it done in time for deadline?" Meigs asked.

"You send *me*," I replied, hoping everyone else ignored the fact I'd never written a feature for them before, not to mention a cover story. Such bravado won me the assignment, but it also put me in a pressure cooker. If I botched this, I would not be likely get the chance to do another.

Now, a few days later, I'm getting a solo tour with the spaceport's general manager, Stu Witt. He drives me around the airfield, eyes darting quickly to read the surroundings the way that even retired pilots do.

Witt was an F-14 Tomcat and FA-18A Hornet jockey, and served as a test pilot on jets like the B1-B and F-16. These days he runs the Mojave Air and Space Port as CEO, watching the runway from his office or his truck.

"We need places like Mojave to be the kindling ground where it's okay to take risks," he says.

Witt knows the history etched in every inch of runway and lurking inside each hangar. He points out the sights: There's where the original Marine Corps barracks stood. This hangar is where John DeLorean's drug smuggling ring operated. Witt stops where Jon and Patricia Sharp store their carbon-fiber kit plane, the Nemesis NXT. The Sharps' previous racer broke so many speed records that the Smithsonian's National Air and Space Museum mounted it as an exhibit. They have a cat who lounges around the hangar.

The flightline is a lineup of rocketeer ambition. Witt points out a hangar where XCOR is building a rocket plane. He curtly nods at buttoned up hangars that host secret projects run by the Defense Department, but even he doesn't know exactly what goes on inside.

Witt is also an ideal spokesman for risk. Not as a dirty word, something to be avoided, but as a necessary adjunct to progress. It's a symptom that can be managed, not something that should be eradicated. Witt once presented a TEDx talk titled "Risk, Adversity and Humanity's Need to Explore." The tenants at Mojave are living examples of the ethos. They are the kind of people who would come to a barren outpost like this and indulge in the kind of aerospace engineering that needs some elbow room.

We pass the composite factory where a legendary firm called Scaled Composites is reinventing airplane design. The huge vents on the side indicate large autoclaves within that bake the composite weave into solid, carefully engineered shapes. The use of this unnaturally strong and lightweight material has enabled designers to take aircraft design in dramatic directions. Witt has watched this firm grow from maverick designers to global leaders in aerospace innovation.

To be sure, Stu Witt did more than just watch. Under his tutelage,

Mojave received its FAA certification as a spaceport in June 2004, the first facility to be licensed in the United States for horizontal launches of reusable spacecraft. By doing so, Witt's tenure at this isolated spaceport would spark a reordering of the entire spaceflight industry.

Stu Witt set the conditions for risk-takers to get their shot, and when they did, it would inspire airports across the nation to take a look at spaceflight. But he was just following in the footsteps laid down in the 1950s by the airport's founder, Dan Sabovich.

THERE WOULD BE NO SPACEPORT IN MOJAVE IF A 1950S ALFALFA farmer never met the fastest man alive. Rancher Dan Sabovich lived in Bakersfield and harbored a serious and ever-growing fascination with aviation. Major Chuck Yeager flew experimental jet planes. The pair were unlikely friends, until you consider that the small population of the entire county (just over 200,000 in 1950) lived in habitable clumps inside Kern County's 8,000 square miles.

In 1947, Yeager loaded himself into a Bell X-1 rocket plane, dropped from a larger airplane, and ignited the rocket. His flight broke the sound barrier for the first time in history, and he accomplished this feat all while suffering from broken ribs caused by a horse riding accident outside the legendary Rancho Oro Verde Fly-Inn Dude Ranch.

This is foundational Mojave aviation canon, as captured in Tom Wolfe's *The Right Stuff*. It has all the hallmarks of the region's rough-hewn, unsung, renegade, and visionary identity. It distills the Mojave spirit into a simple catch phrase, like the name of cologne. "Cowboy Aviation."

Sabovich caught this scent and became enthralled. You couldn't keep him on the farm, especially when his test pilot acquaintances would go out of their way to buzz his house and fields during test flights out of Edwards Air Force Base. Located in eastern Kern County, Edwards is one of the leading military flight test centers in the world. It is one of the most remote, secure locations in the United States, a patch of desert where settlers didn't even think about building homesteads until the twentieth century.

The government needed that kind of space to fly new kinds of aircraft. The vacant land enabled the airbases' footprint to grow in order to accommodate new activity, from bombing runs to hypersonic flight to rocket engine tests. They called it Muroc airbase, taking the name of the town, which itself was named after the Corum family. They spelled the town name backwards, so it wouldn't be confused with another town with their surname.

Spaceflight research began at Muroc in 1944. In 1948, Capt. Glen Edwards died in a test flight of the Northrop YB-49 Flying Wing, an experimental airplane that taught engineers the lessons needed to create the B-2 stealth bomber. A year later, Muroc was renamed Edwards Air Force Base. The seeds of the early research here would grow into the US Air Force Test Pilot School and NASA's Dryden Flight Research Center.

By the 1950s practical rocketry was the hottest game around, and the Mojave desert became the place where the United States tested its newest machinery. The government wanted its own place to study rocketry, lest they fall to the mercy of companies researching it in their own labs and test ranges. So the space age came to Mojave sooner than the rest of the nation.

Dan Sabovich glimpsed the revolution of flight happening at Edwards and wanted to create his own hub. "I knew pilots from Edwards and the flying community from Los Angeles, and knew we needed a civilian flight test center," Sabovich later told the *Valley Press* newspaper.

He knew exactly where to build it, too, just twenty miles east of Edwards. Mining companies opened the airport in Mojave in 1935, and the government had appropriated it during World War II as a Marine auxiliary air station. Pilots destined for combat in Japan received gunnery training in the empty desert wastes. The airfield had since lost its usefulness and the county begrudgingly took it over after the war.

"I can remember when the airport was an albatross around the neck of the Board of Supervisors," Mary Shell, mayor of Bakersfield, wrote in a newspaper column. It was their constant headache, losing money every year since it was given to the county as military surplus after World War II. "Then along came Dan."

In the late 1960s Sabovich tried to buy the airport from Kern County and run it as a private airfield. The effort failed, but it did attract the support of some key figures. Chief among them was Congressman Barry Goldwater, Jr., who eventually helped maneuver the new airport into existence.

Better prepared and backed by powers in Washington, DC, Sabovich finally persuaded the state and Kern County, which owned the property, to turn the airfield into an independent airport district. By 1972 his dream was accomplished: Mojave Airport opened for business, with Dan Sabovich as its general manager.

Today, the world-renowned, civilian-run National Test Pilot School operates from Mojave Airport, a testament to Sabovich's original vision. Tenants can access otherwise reserved military airspace and a supersonic flight corridor, thanks to an agreement with Edwards Air Force Base. There's also a large boneyard of retired civilian aircraft, signs of commercial work being done here.

But Sabovich planted the seeds of the future in 1974, just two years after the airport district was founded, when a young aeronautical engineer from California Polytechnic named Burt Rutan set up the Rutan Aircraft Factory. "Trouble was, I could not afford a hangar," Rutan later recalled to a reporter. "Dan let me have a hangar at no charge. If he had not, I doubt that *Voyager* would have been built."

Rutan would go on to found the company Scaled Composites, a firm that was destined to shake up the world of spaceflight. Scaled's willingness to design groundbreaking airplanes would dovetail nicely with other entrepreneurs making a home in Mojave—the rocket people.

One of the first of these new rocket companies was Rotary Rocket, who arrived in the mid-1990s. Founded by Gary Hudson, the company developed a low-cost, manned, reusable spacecraft, which Hudson calls Roton. Hudson hired Rutan's outfit to help build a sixty-three-foot-high prototype that would launch like a conventional rocket, boosted by a kerosene and liquid oxygen rotary engine. After re-entry, pilots would make helicopter-style landings using nose-mounted rotors. Unfortunately, the company itself never got off the ground, and it finally folded in 2001.

But by the new millennium, other firms were firing engines and blasting rocket-powered craft around the desert. XCOR Aerospace also based at the Mojave Airport, engaged in testing its piloted EZ-rocket as part of an expansive reusable rocket engine and rocket-powered vehicle program. Orbital Sciences Corporation and Interorbital Systems form footholds here, too.

Sabovich ran the airport until 2002, when a stroke caused him to retire. He died three short years later. Stu Witt replaced him, and still refers to his predecessor's risk tolerance as "genius."

Risk is more than a catchphrase or a word bandied about by venture capitalists, and courting it in Mojave can cost lives. Witt knows this, but he doesn't just follow in Sabovich's wake. He doubles down on his predecessor's welcoming attitude toward rocket tests, experimental flight, and emerging industries. His tenants are the most gung-ho pilots and engineers in the nation, filling his ears with fanciful tales of rocket-powered launches from airplanes, of rocket-powered landings, of conquests of the Moon and Mars.

Witt knows what to do—enable his tenants to launch and land suborbital spaceplanes, becoming the first commercial spaceport on the planet. And in 2002, he sets about doing it.

IF THERE'S A PATRON SAINT OF SPACEPORTS, PATRICIA GRACE SMITH (née Jones) should be considered for canonization. She's there to help the upstart airport in Mojave and its renegade tenants risk their lives to further spaceflight and human exploration. She also sets the tone, and the regulations, that allow the private space industry to benefit from and coexist with—often uncomfortably—the United States government.

Smith was not exactly born into aerospace, but it's been around her for her entire life; her father retired from the Air Force and settled in Tuskegee, Alabama, where he worked at a VA cantina. After graduating with a degree in English, she scored a job with the FCC, working on regulations surrounding satellites—an unexpected career move. But she took to the topic, eventually parlaying her knowledge of how the space business works into a job at the Federal Aviation Administration (FAA).

In the mid-1990s, the Department of Transportation (DOT) handled all things regarding commercial spaceflight. The feds had to be involved, since the nationality of the launch operator and the location of the launch determined which country was responsible for any damage that occurred in space. That could leave Uncle Sam on the hook for billions of dollars if some renegade yahoo were to shoot something into space that collided with another satellite or launch something that came crashing down on a populated area like a meteor.

FAA in the mid-1990s was not entrenched in spaceflight, despite the fact that these things sail through the air before ever reaching space. As new launch vehicles and schemes begin to hatch, especially in Mojave, the agency wanted to reclaim this airspace action as part of its domain. Some both in and out of government chafed at the idea. What do these airplane people know about spaceflight, anyway?

For months, Smith worked inside the FAA to prove that the agency could handle the safety mission and enable progress at the same time. This campaign persisted for more than a year while the FAA struggled to take the work from the DOT. In 1998, the FAA won and officially took over the Office of Commercial Space Transportation.

FAA Administrator Jane Garvey named Smith to be the first FAA official to head the office, which everyone called AST. Her title: Associate FAA Administrator for Commercial Space Transportation. Smith and her staff of sixty licensed unmanned rockets[4] hefting satellites or other payloads into space. But she viewed her job responsibility differ-

4. A vehicle's thrust must be greater than lift for the majority of powered flight to be considered a rocket.

ently than the DOT had, becoming an advocate for the new instead of a bureaucrat rear-guarding the old.

An AST-hosted conference in 1999 demonstrated her new approach. Panels at the Space Transportation Forecast Conference included some forward-thinking problems, like space traffic control and the medical impacts of prolonged space travel. Gary Hudson, CEO of the first space company to open in Mojave, the Rotary Rocket Company, joined a panel with famed astronauts Buzz Aldrin and Pete Conrad. "He noted that the Roton will be the only commercial, orbital, piloted launch system in the world when it becomes operational in two years," an FAA recap of the event reads. "Rotary Rocket is proposing the use of the Roton for carrying passengers into space for tourism."

The changes were more than just optics. Behind the scenes, the new FAA office sought to untangle the rules that hampered human spaceflight. The government could have, easily, crippled the effort by adhering to rules that covered unmanned spaceflight. But they did bureaucratic things to clear the way, like gaining official recognition of the pilot as "part of the flight safety system for the vehicle," which freed Scaled to pursue SpaceShipOne's launch.

The rules Smith made followed a simple mandate: Keep the public safe. Those choosing to ride are free to risk their lives. Of course, rocketeers chafed at any controls, even as they appreciated the difference between the DOT and the FAA. Without her tending, orbital plans of Elon Musk and Jeff Bezos could have been crippled at birth. She also jealously guarded AST's jurisdiction over suborbital flight, which after all had most of its mission within the air, not space. Other aviation regulators with less permissive attitudes felt that they should watchdog these flights, but she stymied them.

"She saw it much more clearly and earlier than other people did," Bretton Alexander, a former AST employee who went to work with Amazon founder Jeff Bezos's space company, said in *The New York Times*. "And she plodded away to get there and never let the bureaucratic encumbrances of Washington stop that from happening."

These quiet moves would have far-reaching consequences. The

entire commercial spaceflight industry, from the Martian aspirations of billionaires to spaceplanes carrying tourists, sprang from these new rules and mindset.

"Space is an attitude," Smith said during a lecture at the Arthur C. Clarke Center for Human Imagination in San Diego. "It's a set of capabilities, an acceptance of risk-taking activities to uncover potential breakthroughs and endless possibilities. That is precisely why we love it."

THE MARIAH COUNTRY INN, ABOUT A MILE FROM THE MOJAVE AIRport on State Highway 58, is the nicest hotel in town. Guests can hear airplanes take off, either a buzz or a roar depending on their size, from their rooms. Models of old warbirds dangle from the ceiling, hanging near those hotel chairs that no one ever seems to sit on. Conference rooms, carpets decorated with bewildering geometric patterns.

The people of Mojave gather at 6:00 p.m. on December 10, 2003, to discuss the plan to convert the airport to a spaceport.

Like nearly all government entities, there are a series of rituals that the FAA has baked into the credentialing process, and one of them is the public hearing. There are two ways they go. They can become a forum for complaints, protests, grandstanding, and posturing. Or they can be quiet, rubber-stamp affairs that only validate exactly what the municipal leaders want.

Mojave is not a big place and, really, the airport is the best thing the town has going for it. The cowboy aviation ethos primed the public, and predisposed the residents to accept the designation and whatever it brings.

A spate of groups in town had filed letters, and a handful show up to speak. None speaks out against the spaceport.

Cathy Hansen, a member of the board of directors for the East Kern Airport District, stands. "That's what makes it so exciting. The spacecraft that are launched here can either launch from the ground or be dropped by another aircraft," she says. "I have grandchildren that will come running into the house and say, 'Grandma, look what's up there!' And they get very excited when they see a Burt Rutan–designed airplane. They always know when it's one of Burt's. That is the future."

Others take a more desperate view. "The town just lost a significant revenue source in the building of a highway bypass," said Mojave resident Dan Delong. "I'm looking forward to this new industry that the FAA can support to bring some more jobs back to Mojave, instead of drying up and blowing away."

Burt Rutan's brother Dick, test pilot and black sheep of the family, introduces himself. "I'm a . . . what am I? The Director of . . . I've just been elected to the Board of Directors for the East Kern Airport District," he starts his rambling speech in support of the designation. "This is an emerging industry, a rapidly developing, exciting industry with potential we can't even imagine."

The FAA people promise a yea or nay will come soon and the meeting adjourns. This public meeting is a required event of the all-important environmental impact study, a tall, time-consuming hurdle that every airport must endure on the way to becoming a spaceport. The document analyzes every aspect of the proposed spaceport's activity, from car traffic increases on local roads to the expected sonic booms. The extra particulates in the air, the impact on airspace, the footprint of the launches, the propellant stored on site—everything is wrapped up in a single document.

Reading the 268-page environmental study for Mojave's airport is a good way to appreciate the complexity of launches and the fixations of regulators.

The first priority for the FAA is to make sure that the space operations wouldn't hurt anyone. Again, anyone who hasn't signed up to take the risk, that is. "In terms of impact, for a nominal trajectory, the ground track does not include flights over populated areas," the environmental report says, before adding, "However, in a catastrophic accident, it would be likely that the crew would be seriously injured or killed. At the airport, the onsite fire department could respond, secure the site, but stay clear of the immediate area until the danger of explosions is diminished."

Beyond all the detailed paperwork, studies, and filings obstructing Mojave's spaceport dreams, it seems fitting that the cause of one additional delay was a tortoise. In order to become a spaceport, the airport was required to devise procedures to save endangered turtles that wandered onto the runway. The way they handled the animals needed to match both the US Fish and Wildlife and California Department of Fish and Game protocols.

In 2004, the crumbling wall of bureaucratic resistance to private space begins to fall. The pressure to do something different is building, even then, as the shuttle program's eventual end loomed.

The result comes in the form of legislation. The Commercial Space Launch Amendments Act of 2004, passed by Congress and signed by President George W. Bush in December 2004, gives the FAA's Office of Space Transportation full authority over commercial space launches. Now, a place like Mojave has a one-stop-shop for its spaceflight needs, instead of facing a dizzying outlet mall of FAA divisions, each with differing attitudes about private spaceflight and internal squabbles over who is in charge of what.

The updated law, called the Commercial Space Launch Amendments Act, also establishes an experimental permit for reusable orbital rockets. This opens up an entire spate of spaceport development and a boom in space projects.

"I think it's going to be a wild ride the next twenty years as this industry emerges," Witt told CNN at the time.

In October 2004, the FAA puts its stamp of approval on the airport's plan, and the Mojave Air and Space Port is born. "We got it!" Witt emails the industry trade website Space.com. "It's good to be first."

The same week the spaceport receives its FAA designation, a strange-looking aircraft rolls down the facility's runway. Everything is about to change.

OCTOBER 4, 2004. ALL 27,000 EYES AT AN OLD AIRPORT IN MOJAVE, California are turned up to watch the White Knight carrier plane take off into the desert sky. It heads east, toward the empty desert, but also toward the sun. The gawkers below squint to catch glimpses of the plane as it climbs. At 46,000 feet, the airplane drops its cargo—a suborbital spacecraft called SpaceShipOne.

SpaceShipOne's fuselage is shaped like a bomb, tapered on one end like a bullet and the other narrowing to a five-foot-wide exhaust port. The nose end is mottled with round portholes, seemingly random but in reality carefully placed for the pilot's vision. The spaceplane is shaped like a Y, with twin tail/wing hybrids on each prong of the letter.

The mothership and spaceplane have one immediate goal in mind: Win the $10 million Ansari X Prize. Like some golden age aviation competition, well-heeled enthusiasts offered the prize money to any non-government group who could launch a craft into suborbital space, land it, and reuse it on the same mission within two weeks.

Today's launch is number two. And there's more than just $10 million on the line: People below are betting this flight will kickstart the space tourism industry.

Something about this launch has spoken to the American spirit, and the public interest is unexpectedly intense. The environmental impact study that cleared Mojave for this suborbital space jaunt anticipated a handful of visitors to watch a launch. Nearly 20,000 are piling into Mojave to watch this flight.

Most are in for a rude surprise as the mothership flies straight east, toward the sun. No one thought to share the flight plan with the spectators. Air Force radar operators track the flight to make the official altitude measurements. That's why, when the spaceplane drops from the mothership and ignites its engine, onlookers are staring at the sun. Hundreds of photographers curse as one.

SpaceShipOne, buffeted by wind shear, breaks the sound barrier ten seconds after the rocket ignites, and the spaceplane starts to climb. As the aircraft approaches 200,000 feet, one of the control-surface motors fails, and pilot Brian Binnie switches over to a backup system. But SpaceShipOne's angle is wrong, and that will cost some height, maybe enough to fail the challenge.

The engine is made to burn for eighty-two seconds. The powered flight ends and becomes a high-speed coast, ascending about 3,000 feet up per second. The Air Force radar guys watch the returns come back, ticking closer to the flight's apogee. This is just an exercise in ballistics, a suborbital badminton shot with an invisible sixty-two-mile-high net.

SpaceShipOne crosses the Karman line between space and earth's atmosphere, and keeps going. Brian Binnie takes out a bag of M&M's and lets a few drop. The candies defy gravity and swirl in free fall as the craft reaches apogee, 367,500 feet or sixty-nine miles high. This breaks the rocket plane altitude record set up the road at Edwards by an X-15 in 1963.

What goes up fast comes down hot. To get ready for the trip home, the plane configures its tail, swinging the tail booms and part of the wing upward to produce high amounts of drag. This stabilizes the craft on the way down. Binnie can now angle the spaceplane to aim it for reentry, using cold gas jets to get the right position.

The way down is a fight to control a long, violent free fall. A sonic boom rattles the crowd, still straining to see the spaceplane's return. The deceleration Binnie experiences generated about 5 Gs (five times earth's gravity), pressing him firmly against the seat. Once the worst of that is over, the craft reconfigures again and glides home the final ten miles of altitude.

The crowds cheer as SpaceShipOne appears in the sky, flanked by chase planes taking videos of the historic return. The mothership aircraft, WhiteKnightOne, hasn't even landed yet.

The shepherds of private space are in attendance; Marion Blakey, Administrator of the Federal Aviation Administration, is on hand to watch. "Kitty Hawk, move over," Blakey says. "This was not only a historic flight, the standards of safety that were set here today are going to go on to ensure that there's going to be lots of tourists out there that'll enjoy it."

The praise is breathless, out of control and out of proportion. "Twenty thousand people, who in the early dawn saw SpaceShipOne going like a bat out of hell into space in the blue skies above Mojave, not only experienced an inspiring historical moment, but observed the true beginning of the new space era, a turning point for mankind," writes Derek Webber, at the time the Washington, DC, director of the aerospace consulting firm Spaceport Associates.

Among the throng, Richard Branson stands elated. The British entrepreneur has immediate plans to market space tourism flights to the public under the name Virgin Galactic. The company had announced just the week before that they plan to build a new five-person version of SpaceShipOne and sell tickets for flights. They will call it SpaceShipTwo.

Patricia Grace Smith is present and watches the flight with unabashed excitement. "Is it inspirational?" she later tells Congress. "Absolutely. It's essential. All you had to do was be in the desert in Mojave in 2004 and see the thousands of people who assembled there, young and old, from all over the country, all over the world, to see their eyes light up with the first flight of a private human spaceflight vehicle to know how exciting it is."

This was supposed to be the birth of the space tourism industry. But the impact of Mojave's inland spaceport, and the famous spaceplane it hosted, carries more influence than anticipated. Other airports will take notice of what the desert outpost accomplished. There are about to be a lot more spaceports in the United States.

IN NOVEMBER 2006, THE MOJAVE LEADERSHIP MOVED THE ROTON prototype from the outskirts of the airport to a small park near the port's entrance. Whereas some might have seen the launch vehicle as a monument to failure, Mojave's veteran rocketeers see it as a tribute to audacity.

"When the Roton was relegated to an obscure corner of the airfield, I used to feel like I had wasted three years of my life," Hudson tells me in an interview. "Now that it's been moved, I feel pride when I see it. Prior to Rotary, few people spoke about commercial human spaceflight, only satellite launching. Now everyone does."

Two years after SpaceShipOne's flight, signs of the changing times are everywhere. Tourists visit the ISS by hiring rides on Russian hardware. SpaceX's Falcon 1 flight vehicle stands on a launch pad at Vandenberg Air Force Base.[5] In New Mexico, voters approve a spaceport tax to help fund what will become Spaceport America.

The shifting attitude toward commercial space means changes at Mojave as well. The empty hangars begin to fill as existing companies grow and new ones move in. DARPA and NASA hire firms there to conduct research into sensors and rockets. Engines regularly roar on test stands.

In late July 2007, two things happen that rattle Mojave's sense of independence. First, the aerospace giant Northrop Grumman announces its purchase of Scaled Composites. The firm, a partial owner partnered with Scaled in the Spaceship Company to build the commercial version of the X-prize spacecraft, ups its ownership to 100 percent.

Six days after the deal is announced, a second unexpected event shakes the young spaceport. At the rocket test range in a remote corner of the airfield, Scaled Composites conducts a cold-flow test—a test of plumbing that does not include ignition—on a new engine component for SpaceShipTwo. Three seconds into the test, a pressurized tank of nitrous oxide explodes. Of the seventeen people at the test stand, eleven of them were able to shelter behind a van. The others had lined up by a chain-link fence. Shrapnel from the blast tears through the fence and the Scaled staff standing there. Three are killed, the others injured.

"We had done a lot of these tests with SpaceShipOne," Rutan later says. "We felt it was completely safe."

State and federal government safety regulators descended on Mojave in the wake of the accident. "It was one inspection after another," recalls Bob Rice, the port's operations director. After a six-month investigation, federal and state inspectors were unable to determine the exact cause of the accident.

---

5. This was before their early operations were exiled to a military base deep in the Pacific.

In late 2007, the FAA briefly threatened to rescind Mojave's spaceport license. Instead, after inspecting the facilities, the agency instituted safety-related amendments to the license. The FAA, under Patricia Grace Smith, resisted the temptation to overreach and ceded the investigation to the state of California's OSHA division. "It was not a launch accident. It was not a flight accident. It was not directly related to vehicle performance or passenger involvement," she said in a statement. "The dream of private human spaceflight is in motion and I expect it to keep moving forward."

In January 2008, the state levied $25,000 in fines on Scaled Composites. Chastised, the company continued its work. "We all learned something," Witt says. "It was an eye-opener to see government in action and, in many ways, overreaction."

The tolerant facility that Dan, Patricia, and Stu envision survived the disaster. But this would not be the last time death would stalk the spaceport in Mojave.

IN 2013, STU WITT WENT TO CONGRESS TO PLEAD HIS CASE TO THE House Subcommittee on Space. "My message to you from the high desert is that American engineers and entrepreneurs in Mojave and other places across the country are successfully revolutionizing America's future in space," he testified. "This is a 100 percent good news story. What my Mojave tenants require from elected representatives in Washington is continued permission, and modest encouragement, rather than obstacles."

At the time of his testimony, Mojave Air and Space Port hosted nineteen rocket test sites and seventeen companies performing commercial space research. The firms there were a roll call of ambition and cool-sounding spacecraft: In one hangar, XCOR was building a spaceplane called the Lynx. There is a wooden mockup of the thing in a hangar: it's a rocket plane that designers envisioned shooting into suborbital space and landing on its own, no mothership required.

Then there was Stratolaunch, the next collaboration between Paul

Allen and Burt Rutan. The idea was to construct the biggest plane ever made—with a wingspan the size of a football field—and use it to launch rockets and deliver payloads to orbit. The massive size meant large payloads and multiple rockets could be launched during one Stratolaunch flight.

Allen says the aircraft is being designed to launch cargo and humans into space. In 2011 Stratolaunch Systems signed a twenty-year lease at the spaceport and by 2013 had built its second building at the spaceport. Stu Witt made sure they felt at home by widening the 12,000-foot-long runway to accommodate the airplane's girth.

Popular Mechanics sent aerospace writer Michael Belfiore to profile the creation of the airplane, which people in Mojave had taken to calling "The Roc" after the mythological bird. Somehow, over nearly a decade of planning and development, the project had been kept secret. Belfiore found that Witt, the spaceport king, was in on it. "What brought the Wright Brothers to Kitty Hawk was freedom from the encroachment of the press, freedom from industrial espionage, and a steady breeze," Witt says. "The fact we were able to keep this under wraps for nearly nine years says we still enjoy the elements that took Orville and Wilber to Kitty Hawk."

The message was clear: When it comes to groundbreaking aerospace work, Mojave is still the place to be.

THERE IS NO SHORTAGE OF DREAMERS AT MOJAVE AIR AND SPACEport, and endless reasons for a journalist to visit. I visit again in 2015, finding the flight line seemingly filled to capacity. The aerospace firms here have grown, expanding into empty hangars and adding staff.

It's Thursday morning at Masten Space Systems' main building, a worn out trailer that has stood here since the 1950s. The building looks its age. The Marine Corps operated a training base here in Mojave, and Masten headquarters is located in one of the original aluminum-walled structures they left behind.

The aerospace company's founder David Masten sits at his cheap metal desk, staring intently at his screen. There's a rectangular shape on it, rendered in a CAD file. It's a metal tile, meant to replace the ablative ceramics used during a spacecraft's reentry through atmosphere. Those tiles have to be replaced, so swapping them for tougher materials would make it easier to return a spacecraft to flight. The battle for a reusable spacecraft is fought one component at a time.

But Masten's face is pinched as he wrestles with a way to make the metal seals between the panels hold up under the stress of reentry. Later he'll say that the idea "probably won't work out."

There's an inner geek inside me that equates his angular face, pale eyes, and weather-eroded skin with a Tolkien-style wizard. Of course, the jeans and cheesy t-shirt (decorated with a cartoon of one of his own rocket-powered test vehicles) sort of diminishes the effect. He's a proud, demonstrative man who walked away from what he calls the "golden handcuffs" of Silicon Valley to make rocket-powered craft in the desert. He made himself the Chief Technology Officer of his own company.

NASA, the Department of Defense, and universities have all hired Masten to conduct flight services or design studies. Along the way, Masten routinely demonstrates reusability in rocketcraft by creating landers that can navigate and set down within one inch of target. "I wasn't surprised the revenue came in," Masten says. "I was surprised that it came in before I went under."

The goal of the company that bears his name, Masten says, is to be the provider of transport in the solar system. Other rocketeers dream of building the first manned lander on Mars or the first capsule to bring tourists around the moon. Masten wants to build the spacecraft equivalents of long-haul trucks.

His headquarters, where these cosmic dreams are hatched, is rugged. Close to a dozen desks sit under a water-stained ceiling. The occupants of those desks, mostly thin young men, are scrutinizing CAD renders of various rocket parts. The shared desks hold physics books, sci-fi novels, chemistry reference texts, and 3D printed objects. The metal pencil holders are obviously machined in-house.

The majority of the small staff is gathering in the hangar, reachable through a doorway in the metal building. They form a semicircle to hear a quick safety briefing, the last step before they head onto the spaceport-designated range for a rocket test of their craft, called Xero-B.

Masten has been hired to help a university with an obscure NASA-funded project called Starshade. The theory: that a ground-based telescope can communicate with a spacecraft, guiding it so that it blocks the light of distant stars. By doing this, the telescope can directly detect exoplanets orbiting the stars, in the same way you can better see an airplane when your hand blocks the sun. Today's launch is a dress rehearsal for an impending visit from the Starshade team.

The average age of the Masten staff is in the mid-twenties. Most are male. All are browned by the sun, and either scrawny strong or well-muscled. They are badass nerds. Tan, rugged, resourceful, they're like a clique of heroes from a William Gibson novel. On their off hours they wear t-shirts that say things like "Moon First!"

Most have an understated but real passion for the company and for spaceflight in general. Others, like lead test engineer Ruben Garcia, are devoted advocates. "We can all work elsewhere and get better money and a better retirement plan," he says. "But I don't want to go anywhere else. When I watch the Xero fly today, that's the embodiment of who I am."

The briefing breaks up and everyone separates, knowing what to do from previous experience. Those in the garage ready the Xero-B for transport to the pad. It stands about six feet tall, on four spindly metal legs. There are two large spherical tanks stacked atop each other—one containing isopropyl alcohol fuel, the other empty but ready to hold cryogenically chilled liquid oxygen. It's against FAA and airport regulations—and an all-around bad practice—to load an oxidizer and fuel anywhere but on the pad.

Another ring of tanks circle the waist of the craft. These are canisters of helium, used for the RCS thrusters that help orient the craft. The engine at the bottom is also gimbaled to give the insectoid, threadbare craft the ability to steer.

Masten's crew operate several landing pads and have created rocket-powered craft that can fly hundreds of feet high and make precision landings on pads nine hundred feet away. Today's test is not as dramatic—the rocket will fire more than half a minute but the craft will remain on a tether. A critical point of the exercise is to hone the ground procedures for a flawless launch and quick turnaround to prepare the test craft for another launch. The data from the engine should not hold any surprises; clients expect test bed aircraft to act predictably.

Reusable rockets are about more than the hardware. Without an experienced launch crew that can handle the fast pace, the hardware would be useless. The devil is in the details: the personnel manifest filed to the FAA and airport fire department, the safety equipment kept at the ready, the tools needed for the job, the creation of the flight plan filed with the FAA.

Handling the hardware is a well-practiced drill as well. A wheeled utility vehicle with a telescopic arm creeps to the hangar door and the staff quickly attaches harnesses to the Xero-B. The utility vehicle's arm raises the rocketcraft and places it in the bed of a dented, blue Dodge Ram 2500. Crewmen promptly lash it in place with canvas straps and the truck rolls off again.

At the launch site, two interns are sweeping sand and dirt from the concrete launch pad. Everyone is wearing color-coded long sleeve shirts—white for workers, blue for supervisors, and red for safety specialists. There are five white shirts on the pad today, a larger than usual number by three, but this is a sign of growth. These are new employees sweating under the desert sun to see the process up close.

A pair of white shirts set up lights and don headlamps. The Starshade tests will occur at night, and the team is doing a high-fidelity run-through even though the sun is shining.

Any professional rocket launch is a series of checklists and countdowns. Cell phones are ordered disabled, for fear of interfering with the telemetry data streaming to the command bunker, two hundred feet from the pad. Inside the bunker, systems engineer Jeff Gibson cycles the valves, triggering them one by one as workers in headsets report when they hear clicks. "RCS-2." *Click*. "Good." RCS-3." *Click*. "Good."

Gibson designed the trajectory, and he is the one who knows what the flight looks like via the telemetry, pressure, and other data that streams onto his laptop. A pale young man with a ruddy beard and scarred hands, he admits to some sentimentality toward the Masten rocket landers. "You work with these things day in and day out," one says of the Xero-B. "You really do start to care about them."

It's time to fill the liquid oxygen tanks with oxidizers. Helmets are donned, face shields lowered. Hazardous ops have begun—the explosive reaction between fuel and oxidizer makes rockets move, when controlled. If the chemicals react when they are not supposed to, the energy bursts at once, dangerously. The checklists and protective gear is meant to keep accidents rare and harmless.

The Masten crew connects silvery hoses and turns wrenches, and a loud hiss fills the air as the liquid oxygen tank fills. After a break, during which the turbulent liquid settles, the radio crackles. "Top off has begun," a voice says. A roar overhead drowns out the seething sound of the transfer—an Aermacchi or F-4 jet cruising past, probably from the flight test school here.

The final countdown is not heard on the radio—all frequencies are reserved for the pilot, who can abort the test at the last second if need be. So the intense roar that accompanies the cone of red-orange flame under the Xero is a shock, no matter how expected. The PVC pipes under its legs skitter away into the dirt, the harness going slack as the craft rises. Tens of seconds go by. Water, long ago trapped in the concrete of the launch pad when it was wet, evaporates so quickly that small chunks of material break away under the fiery assault.

The rocket lowers and the noise stops—the flight is over. Data has streamed from the pad to the command center, where Jeff can quickly see that the flight profile has matched what he planned. Lessons learned are aimed more at the people than the vehicle—bring more plastic ties, make sure the headlamps are ready, and so on.

The crew vents excess liquid oxygen, loads the Xero-B back onto the truck, and sets back to headquarters. There, Dave Masten is seated at his desk, again—or maybe still—pondering that metal tile CAD file on his screen.

STU WITT RETIRED AS SPACEPORT CEO IN 2016, AND A GROUP OF aerospace people gathered to pay him homage at a ceremony. Maybe most strikingly, Rep. Kevin McCarthy (R-Bakersfield) came from Washington, DC, bearing a signed, framed copy of the US Commercial Space Launch Competitiveness Act, passed in 2015. "I may be the author, McCarthy says. "But the fingerprints all over the Space Act are his."

Witt left aerospace a better place than he found it. More importantly, he left Mojave as a healthy, experimental spaceport, still a safe place for development of suborbital spacecraft.

It's a sad fact that none of the spacecraft born here plan on staying. That makes it tempting to see Mojave as the small town music venue that watches its biggest successes move away and its failures fade to obscurity. Don't. This place is positioned at the cutting edge, and there are always customers willing to go there to try their hand at aerospace glory. Mojave is the incubator for wild ideas, and it will continue to attract that kind of aerospace project.

Besides pride, there's a stable business here. It's a spaceport's biggest secret: They don't make money from the operation of flying spacecraft around. Renting hangar space to pilots, providing room to test engines, serving food at the Voyager Cafe, constructing wind turbines, rehabilitating airplanes, and running the test pilot school—these things make money. The flights of experimental aircraft here, especially the first-evers and record-breakers, bring attention to the spaceport and show that the aerospace companies are busy. But making money from flying spacecraft as a main business is much riskier.

Airports don't make money from landing and parking fees for airplanes. Concession stands, car parking, retail stores, rental car counters, and the rest balance the books. According to the Airports Council International-North American Concessions Benchmarking Survey, non-aeronautical revenue reached 45 percent of total operating income in 2013, worth $8.2 billion. The airplanes themselves only generated 30 percent. Those trends have only gotten more dramatic since that study was released in 2014.

In the years since, more states and municipalities have organized to create their own spaceports. They are luring spaceflight companies, many of which did development time in the Mojave Desert, to anchor their operations. But what they need are clients who can treat the spaceport like an industrial park—renting space—until the spacecraft are ready to fly.

Mojave has what they all want: clients working on projects that are not dependent on any single spacecraft program transforming into an actual paying business. That keeps Mojave on the map, while the newer spaceports must build places in hopes the companies will arrive.

It would be trite to say that the biggest asset here is the people. The feats done here are legendary, and that feeling pervades even the most routine parts of the work they perform. Lunch at the Voyager Cafe isn't just time spent at a greasy spoon; it's a chance to gawk at the engineers cracking spaceflight open to industry. Imagine a time traveler who could sit with the guys who built the *Titanic*. Put yourself in the shoes of someone who could dine near the team who constructed the Spruce Goose for Howard Hughes.

That's the feeling visitors get at Mojave Air and Space Port. People who work here appreciate this, but see it differently. To understand their Mojave, you need to go outside and look up at that fabulous airspace. Just like a shoreline encourages the human spirit to jump on a boat and explore, the empty sky over Mojave invites those below. And the airport that Dan Sabovich built is still ready to host those who want to construct a vehicle and set sail. How far is just a matter of their ambitions.

CHAPTER 4
DESERT HUBRIS

THE CITY OF TRUTH OR CONSEQUENCES HAS A HISTORY OF going overboard for the next great thing. At first it was the hot springs, heated mineral-rich water bubbling from underground that was said to have regenerative properties. When residents formed a town in 1916, they called it Hot Springs. It served as good marketing; by World War II more than forty spas operate in the area.

Alas, the idea of basing an entire tourist industry on hot springs isn't exclusive to this isolated patch of New Mexico. Inconveniently, there are other places in Arkansas, Montana, and California also called Hot Springs.

New Mexico truly feels remote. No one comes here without a reason. Someone in Roswell, New Mexico, once told me, "This is the state where the US government hides things. There's no one out here to see it." In 1950, those things included atomic bomber airfields, experimental missile ranges, and warplane training bases.

Making a trek out to Hot Springs means long hours on hot, sparsely populated roads, a much less appealing drive than to those other Hot Springs destinations. The state needed to do something to seize the public's attention. So the State Tourist Bureau came up with a scheme. Staff heard about the offer by game show Truth or Consequences to stage a live broadcast from any city that adopts the show's name. They prompted city officials to set up a vote and, in March 1950, the will of the miniscule population is tallied: 1294 to 295 in favor of name change. Thus, Truth or Consequences, New Mexico, is born with a vote and a publicity stunt.

In the fifty-eight years since that vote, the town hasn't exactly thrived. In 2008, only ten spas remained open. The population of retirees and desert dwellers scraped by on tourist dollars, with young visitors from California drawn by the natural beauty of the region and the town's iconoclastic vibe. But the biggest source of tourists—those travelling around their own state—is still lacking in New Mexico.

But there was another scheme, and another vote, looming. This time, a billionaire from out of state had his eyes on a patch of desert outside of town. He wanted to build a spaceport there, and the county was expected to pass a tax increase to help pay for it. The optics were bad: taxing residents of a poor area (mean annual income of $23,000) to build a spaceport for a billionaire's business plan. And that plan revolved around launching rich people into space, for fun.

Debate erupted, and another referendum was called. In 2008, with 31,022 votes tallied, residents decided to make another leap of faith and authorize the $228 million spaceport. Within months, trucks and heavy equipment started rumbling through the town, heading east along Highway 51. A local woman is arrested for blocking one in the center of a road. Those crews built a beautiful spaceport. And it's been sitting in the desert outside Truth or Consequences, virtually unused, for more than five years.

It's now a monument, but no one can yet say what it stands for. Is it a Greek tragedy warning about the limits of a single man's ego? A testament to optimism, to sacrificing now to secure for a better future? Is it a lesson on how local governments are eager to be fleeced to attract new business? Or is the spaceport just a reminder that Truth or Consequences is still willing to roll the dice to get some recognition?

The answers remain elusive because the final history of Spaceport America remains in the future. Space launches, suborbital and maybe even higher, could still rise from Spaceport America. It could become the hub of a new space tourism industry. It could spur visitors and jobs, and bring young people to the state. Yet anything less than these rosy outcomes will mean the spaceport was not worth the effort.

Whether it proves to be viable in the end, the unsteady rise of this desert facility has already cast a pall over other spaceport projects, even those with entirely different histories and missions. This project was the general public's first exposure to the private space movement, and when it suffers so does the larger industry. The Spaceport America experience is a big part of the reason why "spaceport" has become synonymous in some circles with "taxpayer-sponsored boondoggle." So in a wider sense, the damage has already been done.

I decide to visit Spaceport America at the start of 2017, at the last possible time before this book is due to the publisher. I've been waiting for something to fly from New Mexico, all this time, as a dramatic scene. Like everyone else, I've been disappointed.

Still, I try to keep an open mind. There's nowhere quite like Spaceport America. There's got to be more to it than a temple of unbridled ambition. I want to understand how this happened, and then take a look at the place up close.

SPACEPORT AMERICA OFFICIALLY OPENED ON OCTOBER 18, 2011, and to mark the occasion Sir Richard Branson pulled a stunt. He and his two children clasped themselves into harnesses and, flanked by a troupe dangling beside them, performed a dance routine on the windows of the Norman Foster–designed terminal and hangar. Nearly 1,000 people gathered in front of the sleek building to watch the show. The Bransons wore black, the dancers red.

Branson is a professional showoff. Branson's team will buy an existing operation and market it under the well-known Virgin name. Branson himself doesn't invest much of his own money, preferring to bring in a slate of wealthy partners to pony up the cash. He calls it "branded venture capital." The Virgin Group empire is worth an estimated $5 billion and includes a disparate collection of ventures, including an international airline, publishing company, cosmetics retail and a bridal-wear shop. The V in the name swoops like a check mark, which Branson calls his company's "seal of approval."

"I don't go into ventures to make a fortune," he once said. "I do it because I'm not satisfied with the way others are doing business."

The brash businessman sells spaceflight with zeal. He speaks of launching tens of thousands of customers into space. "Let's go twenty years forward," he tells the BBC. "If all of this goes to plan, I hope that we will have a hotel in space; and in that hotel I hope we will have small spaceships that can go around the Moon on excursions."

To Branson, the explosion that killed three people at Mojave in July 2007 was a just bump in the road. *The Associated Press* quoted a memo from Stephen Attenborough, Virgin Galactic's astronaut liaison, at the time saying the impact of the Mojave disaster was "minimal" and that commercial flights were still expected in late 2009 or 2010. The company was even confident enough to sell tickets, each with a $20,000 deposit. They made sure to tell the media which celebrities had signed up for the trip, too.

To conquer the cosmos, Virgin Galactic first needs to carve out a home base. Plans form in 2004, when New Mexico Governor Bill Richardson and his team of economic developers embraced the concept and offered Virgin incentives to build a custom-made spaceport in the middle of barren land in between the San Andres and the Cabello mountain ranges. In 2005, the state legislature established the "Spaceport America Regional Spaceport District." The project was on its way.

In many ways, the location is ideal for an inland spaceport. The spaceport is seventy-five miles from Las Cruces and 150 from Albuquerque. It's even thirty miles or so outside Truth or Consequences, with its population of just over 5,000 people.

The spaceport's land may seem empty, but it has enviable neighbors: a military base with some of the best airspace in the world—the White Sands Missile Range. With a simple phone call, the massive area of controlled airspace over an empty desert is available for spaceflight. Drop an empty booster or crash a rocket-powered craft out here, and no one will complain. That makes this spaceport suitable for vertical launches as well as spaceplane flights, and Spaceport America has the permits and pads to handle both.

"There was nothing out here," says Dr. Bill Gutman, the spaceport's director of operations. "There was no water, so we had to create this well field. We had to bring in electrical connectivity with the outside world."

Virgin Galactic would be the anchor tenant, and its operations would pay off the startup cost as others come to the party. In 2009, they signed a twenty-year lease. The budget to build the spaceport would fluctuate over the years, but eventually the total cost settled at $209 million. The state paid for two-thirds of the cost, but the locals had to buy in as well.

In April 2008, Sierra County approved the release of more than $40 million in construction bonds for the spaceport and infrastructure to support it. Why would they agree to such a thing?

The mayor in 2017, Steve Green, is not too surprised that a population dominated by senior citizens living on fixed incomes would support a spaceport. The amount is small, he says, at one-quarter of one percent sales tax in an area already spared from payments. "There's no sales tax on food or prescription drugs," he tells me. "That makes it easier." To sweeten the deal, 25 percent of the amount raised goes to local education.

For local officials and business, the allure of new visitors was the critical factor. "The selling point was the job creation from tourism," Green says. "We are a tourism-based economy. We made sure to have the Spaceport America visitor's center located here in town."

Even that, though, was met with resistance. Locals complained about a senior recreation and community center being moved out of the building without warning. The night before the ribbon cutting, dissenters on the city council attempted to pass an ordinance demanding the seniors get the building back. It failed. Protestors and counter-protestors showed up, appropriately sized to the town: a dozen people with signs, arguing in the street.

By October 2010, the runway was complete and the terminal building under construction. In October 2011, Branson told the British newspaper *The Telegraph* that flight tests should extend through 2012, and commercial launches will begin in 2013. "'We're very, very close now, with the spaceport finished, with the mothership finished, with the final tests going on, to starting commercial spaceship travel."

Despite the hype, Spaceport America tenants beyond Virgin proved hard to find. The new governor, Susana Martinez, cast around for private space types to move in and subsequently landed SpaceX as a customer in 2013. At the time, Musk's firm was eager to test hardware to enable empty fuel boosters to fly back for reuse. Such a system could shave tens of millions from launch costs. In time, the deal would prove to be a disappointment when SpaceX conducted this testing in Texas instead.

UP Aerospace have been headquartered at Spaceport America from the start. They use the vertical launch range to loft single-stage, twenty-foot-long rockets dozens of miles into the sky. Humble as these suborbital operations may be in the grand scheme of things, they are at least a launch company with actual hardware. They actually have the highest altitude of anything flown at Spaceport America, at seventy-five miles.

But really, all bets are on Virgin Galactic to get its operation flying. But in 2014 problems with SpaceShipTwo's engine became public. They fired their co-developer, Sierra Nevada, and moved the project in-house.

These delays extended a string of steady losses at the spaceport. Since inception, Spaceport America has run an annual deficit of about $500,000. Virgin's rent doesn't even cover the $3 million cost of fully staffed spaceport fire department. "Being purpose-built will only get you so far," Mayor Green says. "There are other spaceports out there, competing."

The Mojave accident in 2007 tested the patience of those with vested interests in town. Mark Bleth operates a tour business that runs tourists from town to Spaceport America. He started the tours in 2010 when there was "no roof and dirt floors" on the main building. He called these "hardhat tours." As much of a focus the spaceport has on tourism, the reality is that the place was made to develop proprietary and even militarily secretive projects. Guests are sometimes not really welcome. The long, winding road to Spaceport America, which often washes out in the rain, ends at a piece of sculpture and a security gate. No unannounced visitors are allowed.

"We had a captive audience. The only way to see the spaceport was through us," Bleth now says. Access was usually not a problem. After all, "we didn't have any missions to disrupt our schedule."

Bleth envisioned fleets of tour buses on launch days, events like the ones Mojave enjoyed, with thousands of people eager to witness history. He also became a true believer. "It was kind of enlightening when I took the tours out," he says. "It wasn't Virgin Galactic or Spaceport America specifically. I learned why we should care about the coming of the second space age. It's not just about joyrides for millionaires."

But as the years passed and the spaceport remained stagnant, Bleth began to give up hope. He eventually quit the tour business and took a new job running Truth or Consequences Municipal Airport. "I never did make any money on it," he says in 2017. "I believe in the second Space Age but I can't affect that change."

For Green, the activity at the spaceport has been positive, despite the delays. As the spaceport rents itself out to any who want it—film crews for car advertisements, drone conferences, space industry companies and the occasional defense contractor—the only place to stay nearby is Truth or Consequences. "Have we got a return on the investment? Absolutely," he says. "Is it the return on investment we like to receive? Absolutely not."

Spaceport America's story illustrates the first dilemma of any spaceport—no one wants to build one without a prospective tenant who makes it worth the money and effort. But that places an emerging spaceport itself in the uncomfortable position of fronting lots of money and effort on a venture that may never fly. Taking a commercial space customer also means committing to a partner working on experimental hardware in a young, unproven market.

Patricia Hynes, the director of the New Mexico Space Grant Consortium since 1998 and a major supporter of Spaceport America, sees lessons in their mistakes. "Don't get too far ahead of the market; that's one big lesson learned," she warned aerospace conference attendees at a 2016 panel on spaceports. "We got way out in front."

THE ROCKET PLANE ROARS TO LIFE OVER THE MOJAVE AIR AND SPACE Port on Halloween in 2014 . Any day when a flight happens is a big one here, especially if the spacecraft hails from the hangars of the legendary Scaled Composites.

Scaled has been contracted by Virgin Galactic, through a fully owned subsidiary called the Spaceship Company, to build the upsized versions of the vehicles that won the Ansari X Prize. Instead of just one pilot, the new system will have a crew of two and a cargo of six paying tourists. The company is lagging behind Virgin Galactic's ambitious schedule, running the boasts of Virgin founder Richard Branson against the harsh reality of engineering. Making a flight-ready rocket plane that can reach space after being dropped from a mothership is not as easy as he figures.

Today's test represents welcome progress. The spaceplane, SpaceShipTwo, won't leave the atmosphere today, but the flight test is a happy reminder that the company is edging closer to taking paying customers into suborbital space. There are seven hundred people who have bought tickets for $250,000 each.

There is an entire spaceport built for them in New Mexico, and the taxpayers footing the bill for the bond helping to pay for it are growing impatient.

SpaceShipTwo's engine is flickering a red-orange flame, reflected in the shiny metal rudders. These are the "feathers" that make this spaceplane so unique—they rotate in flight to increase drag and stability when the vehicle returns to Earth.

Pilot Peter Siebold calls out the increasing speed to ground control and, in the right hand seat, sits copilot Michael Alsbury. The crew hasn't flown under rocket power in a year.

By the time the speeding craft gets close to breaking the sound barrier, the aerodynamic pressure building under the tail boom is enough to push the feathers into their upright position. There are locks to keep them in place.

The pilots' normal procedure is to wait until the craft reaches Mach 1.4 when the craft's position has eased the force on the tail. But today, Alsbury flicks the cockpit latches that unlock the feathers about fourteen seconds before he is supposed to, when the craft is moving at Mach .09.

All the systems on SpaceShipTwo are manually operated. Like a hand-cranked window in an old car, the lack of automation is considered to be more reliable than something automated or digitized. Human error isn't considered.

The flight goes awry quickly as the tail boom pivots open, shearing the wings and tearing the entire rocket plane to pieces.

Debris rains down across miles of the Mojave Desert. There's a flash of red, a single speck in the sky. This is Siebold, injured and oxygen-deprived, descending under a parachute. His shoulder is dislocated, his eyes speared with bits of metal and fiberglass. Thankfully, the pilot suffers no permanent damage to his eyes.

A National Transportation Safety Board (NTSB) report contains details of his escape: "The last thing he recalled in SS2 was a very violent, large pitch-up with high Gs, and grunting noises. He heard a loud bang followed very quickly by signs of a rapid cabin depressurization. In the background he heard the sound of 'paper fluttering in the wind,' which he believed was the sound of pieces of the cabin coming apart. There was then a period when he had no recollection, which he attributed to 'g-lock' due to the unexpected onset of high Gs for which he was not prepared."

Alsbury is found dead, strapped in his seat.

Branson speaks to the press in the aftermath. "In testing the boundaries of human capabilities and technologies, we are standing on the shoulders of giants," he said. "Yesterday, we fell short."

The NTSB announces their findings months later. They cite Alsbury's mistake, and Virgin Galactic seems happy to blame the crash on a human rather than the vehicle. But Scaled "set the stage" for the accident through its "failure to consider and protect against the possibility that a single human error could result in a catastrophic hazard to the SpaceShipTwo vehicle," the board's findings say.

NTSB board member Robert Sumwalt adds, "The fact is, a mistake was made here, but the mistake is often a symptom of a flawed system."

Think about it: If there's a red button in your car that, if you press it at the wrong time, will result in the wheels coming off, you'd think that the car dealer would warn you against pressing it at the wrong time. A better move would be to create a safeguard to ensure that such an error, one with catastrophic consequences, couldn't happen. Even professionals make mistakes—that is the very essence of the checklist-heavy world of aviation.

"Commercial spaceflight stands on the verge of becoming a reality," scolds NTSB chairman Christopher Hart. "But the success of commercial space travel depends on the safety of commercial space travel, at the level of every operator and every crew."

Branson creates a video to lament the dead and reassure customers. He seems grimly content to say his vehicle works "as it was meant to" and heap all the blame on human error. "With the investigation completed, Virgin Galactic can now focus fully on the future with a clean bill of health and a strengthened resolve," Branson says in the video.

It will be years before a version of SpaceShipTwo flies again.

The persistent but routine delays are bad enough; an accident—what the mayor of Truth or Consequences describes to me as "when shit hit the fan"—is certainly an unwelcome development.

"From my point of view, most of the people who voted for the bond, and it was a majority, still support Spaceport America," says Mayor Green. "Are there naysayers walking around town? Oh yes."

THE ROAD TO SPACEPORT AMERICA IS AS DRAMATIC AND WINDING as its development. The first Spanish traders who dared cross this expanse of desert in the 1600s called it Jornada del Muerto, "the journey of dead men." The endless sequence of scrub brush is broken by *arroyos*, places where the road dips into small canyons and the roadside vanishes into steep, vehicle-swallowing inclines.

The road comes to a sharp right-angle turn near a handful of brown, one-story buildings. What at first looks like a small country general store is actually an entrance to the 362,885-acre Armendaris Ranch, owned by mogul Ted Turner. Inside is pristine desert terrain for endangered species, lava tubes brimming with bat colonies, and a Spanish mine dating back to 1658. I remember something the mayor mentioned: In 2013, Turner bought the Sierra Grande Lodge and Spa, partially in hopes to provide luxurious accommodations to the well-heeled Virgin Galactic customers. "There are three billionaires doing business in the county," Green says during my 2017 visit. "How many places can say that?"

I have a good guide waiting for me in Dr. Bill Gutman. He's been involved with Spaceport America well before its inception. He's been the resident genius around these parts, working as a research physicist at the New Mexico State University's Physical Science Laboratory in Las Cruces. He looks the part of a desert-worn, lanky, and bespectacled scientist, the kind that haunts universities across the American west. But the proximity to Spaceport America changed his life, starting in the 1990s.

At NMSU, he studies the influence that the altitude, 4,600 feet above sea level, has on launches. It's not just about being thousands of feet closer to space, although that certainly helps. The reduced atmospheric pressure means there is less drag, meaning less fuel or greater payloads. He calculates 6,000 extra pounds of savings, compared to Cape Canaveral.

The state has called on him to be a technical advisor numerous times over the years, performing risk assessments of space and aviation systems to help coordinate the effort to build a spaceport. He eventually co-wrote the spaceport's FAA application, which was accepted in December 2008.

Now, nearly a decade since that moment, I'm eager to hear Gutman's take on the spaceport. He meets me at the gate and escorts me inside Spaceport America. From the outside, the domed terminal-hangar building looks like a landed UFO. The facility appears as empty as advertised. This is where the customers would both check in and board the spaceplane. The runway is fairly clean, the massive metal doors of the hangar sealed shut. The whole place seems lifeless except for two plants near the front door, in the future to be used by paying astronaut tourists but locked today.

I see the infamous fire station. It's sleek and domed, like the main building. It looks more expensive than it has to be. Gutman says that it is. "They were just going to build a plain rectangular building," he says.

So far, the spaceport seems attractive and frivolous. But there's something going on. For the operations manager at an empty spaceport, Gutman sure seems busy today. As we drive around the facility, spaceport staff intercept to deliver concerns. There's a large delivery coming into Spaceport America, and they need a place to store it.

Tests out here are dangerous, secretive, or both, so this is not as easy as it sounds. Explosives must be handled under Bureau of Alcohol, Tobacco, Firearms and Explosives (ATF) guidelines, and security personnel must be on hand to guard proprietary hardware. Spaceport problems, I think at first. But Gutman seems awfully distracted. There are three pallets coming in by truck, and one weighs five hundred pounds.

"We have an event coming up," Gutman says lamely. I tell him that nothing I write will appear for months, so whatever secret he is harboring is safe with me until then. He loosens up and admits what's arriving: a high-fidelity stand-in for Boeing's manned capsule, the CST-1000 Starliner. This is the spacecraft designed to bring astronauts to the International Space Station, built and operated by Boeing on behalf of NASA.

I'm surprised to hear this, and indeed I'm *supposed* to be surprised. Boeing has never tested its capsule here before, at least publicly, and, for whatever reason the company doesn't want people to know about it this time. Spaceport America representatives have hinted at Boeing's involvement, but disclosed no specifics.

In the next few days the company will drop the capsule mockup from some altitude and fall under four parachutes. (In 2012, Boeing did a similar test over a dry lakebed in Nevada, dropping the capsule from 11,000 feet.) The capsule is designed with an extra mechanism to break its fall—a pair of airbags that inflate with a mix of nitrogen and oxygen gas. Today, there is some new iteration of the landing system at Spaceport America.

Gutman can defend Spaceport America by pointing out the work being done here aside from space tourism. "If you look at spaceports across the country, we are one of the few with customers," he says. "This is not a boondoggle."

We retreat to the mission control building inside the terminal. It's not an air traffic control tower—there is none here, and the spaceport rents radar-tracking services from the military—but sort of a conference room with big screens, communications across the facility and wide windows that overlook the full length of the runway. (The runway had to be lengthened to support Virgin's spacecraft, which kept getting heavier as the design matured. The heavier the aircraft, the more room it needs to roll to a stop during a landing.) The view shows just how close, and closely tied, the spaceport is to White Sands Missile Range. A mountain range, ten miles distant, marks the start of the military facility. It can handle a more brisk operational tempo than the airspace around Mojave, Gutman says. If only the customers existed to take advantage of the opportunity.

The windows also provide a view of the vertical launch range, five miles away. "White Sands is big, one hundred miles long," Gutman says. "But the area is only forty miles wide. Boosters land two hundred to six hundred miles away downrange."

SpaceX's involvement there could have been the harbinger of more business, but it turned into another bust. Elon Musk's quest to land empty fuel tanks and reuse them once had a home here. When

SpaceX moved in to test reusable boosters, launching craft and setting them down under rocket power like a 1950s sci-fi movie, Spaceport America officials saw a way to beat the inland geography. Instead of crashing empty fuel tanks hundreds of miles away, they would be coming back for pinpoint landings. With such reusable boosters, Spaceport America could one day launch heavier rockets.

But just like that, SpaceX moved its operation to Texas. Instead of testing the returned boosters in New Mexico, the company skipped a step and sent the boosters from the engine test center in Texas directly to Florida, where they landed the boosters after actual launches. These attempts were televised and, after a few explosive failures, SpaceX succeeds.

"They thought it would take a lot more testing to land those boosters," Gutman says. "We were overtaken by events."

Gutman is a sincere and earnest spaceport official. His retirement job provides a firsthand look at the commercial space industry, albeit one moving too slowly for his taste. He feels the excitement after the X-Prize obscured the difficulty of engineering Virgin Galactic's launch hardware. "It's the same technology, but not quite," he says. "They found that it was harder job to scale it up than they imagined."

Virgin Galactic and the New Mexico spaceport believers are hoping that a payoff for their forced patience may finally be here. In 2017, Branson breaks a hard-learned rule about putting dates to flight milestones. (The thinking is, if they don't publicly commit to a date, they can't be accused of suffering more delays. I find the argument silly, since there are investors, engineers, and launch customers who certainly have dates marked on calendars.) But Branson sounded a very positive tone, supported by continued testing.

"I think I'd be very disappointed if we're not into space with a test flight by the end of the year and I'm not into space myself next year and the program isn't well underway by the end of next year," Branson told the *Daily Telegraph* newspaper. "After whatever it is—twelve years of hard work—we're nearly there."

Virgin Galactic is also correcting a mistake as it embraces satellite launches. The company did commercial space a disservice by focusing on high-end space tourism as its primary market. This move, with movie stars and pop icons signing up for flights, made new space easy to dismiss as a frivolous vanity project, by and for the wealthy. Everything about Virgin stunk of elitism, and primed people to gloat over the failed expectations.

In March 2017, Branson announced the creation of Virgin Orbit, dedicated to satellite launches. "It has been my long held dream to open access to space," Branson said. "We have been striving to do that by manufacturing vehicles of the future, enabling the small satellite revolution, and preparing commercial space flight for many more humans to reach space and see our home planet. I'm thrilled that our small satellite launch service has now progressed to the point it merits the formation of its own company."

The pivot began years ago, as Virgin engineers designed and tested its LauncherOne system, using the airplanes originally created to loft rich tourists. For extra gravitas, the company snatched up Dan Hart, Boeing's Vice President of Government Satellite Systems, to head Virgin Orbit. For three decades he's been responsible for all of the aerospace giant's satellite programs for the US and other governments. The company is spouting some familiar, beneficent language about its business model. "To me, the Virgin brand is about making life on Earth better," Hart said when the Virgin Orbit formed. "We are going to fulfill that purpose by accessing low Earth orbit to connect billions of people and enabling valuable applications of data from space."

That means true orbital spacecraft could launch from Spaceport America. Gutman expects business to grow as satellites get smaller and air-launched rockets become more common. "When we break into orbital flight, the vehicles that launch here will be small ones," he says.

Gutman's phone keeps chirping as the delivery time arrives. The one-on-one tour is finished. He drives me back to the gate, pulling past a work crew manning a large green crane. A gumdrop shape sits on a pallet: the Boeing capsule, preparing for a short stay in a secure hangar, a brief flight, and a long, long drop.

When I wish Gutman good luck, I mean it. I can still see the top of the crane from my parked car, but the rest of the scene is obscured even under binoculars. If you didn't know it was happening, Spaceport America would look like the ghost town everyone describes.

CHAPTER 5

WILD HORSES AT THE PILOTLESS RESEARCH CENTER

DALE NASH TAKES HIS POSITION TO SPEAK TO A GAGGLE OF SPACE press, corporate VIPs, and social media guests. No one is listening too closely.

About fifty feet behind him stands a stark, white 130-foot-tall Minotaur V rocket, tucked inside a rectangular building but visible through massive open doors. The white structure, called a gantry, will lean drunkenly away from the rocket during the last seconds of the countdown.

Everyone is staring at the rocket, not at the executive director of the Virginia Commercial Space Flight Authority (VCSFA), which owns the launch pad upon which the rocket sits. Members of his staff weave through the crowd distributing information sheets and stickers. He's wearing a blue polo shirt, but instead of a horseman it features large block letters spelling MARS.

It's a proud day for the Mid-Atlantic Regional Spaceport. Don't let the cheesy acronym fool you—nothing launched here has ever been to another planet. In fact, this 2013 launch represents the first deep space launch for this oft-forgotten spaceport. The day after

Nash's speech—September 5—the gantry will ease away from the pad to enable the launch of the Lunar Atmosphere and Dust Environment Explorer.

This underdog spaceport is in the midst of a moonshot.

The launch facility is located southeast of Washington, DC, on the ragged coastline of northern Virginia. Roads and cities squat on comfortable swatches of land spared by the Atlantic. It's a fertile area with a long history of Native American fishing settlements. The modern era has transformed it into a quaint tourist hub, with a network of beach communities and a thriving ecosystem of restaurants, hotels, ice cream parlors, condos, and beach gear supply stores.

The big draw, beyond the water, is the presence of wild horses. The horses of Assateague Island are feral creatures that have lived in the pine forests and salt marshes of Virginia and Maryland since the Spanish ditched them here after a shipwreck in the 1600s. They do so well there that their population is hard to sustain. Every year since 1924, more than one hundred mares are rounded up, their foals sold off, and the horses returned to the island. This wildlife management effort, true to the area's economy, has become a highlight of the summer tourism season.

Driving south to Wallops Island, the first visible sign of anything space-related is the massive satellite dishes that grow in lush fields like monstrous mushrooms. The National Oceanic and Atmospheric Administration (NOAA) uses these dishes to process data from the Geostationary Operational Environmental Satellite (GOES) system. The weather images broadcast on network television come from this data.

These NOAA structures are not the only strange dishes and antenna that sprout on Wallops Island. When rockets lift off, radar and telemetry data stream from the rocket into dozens of NASA receivers that are speckled across the coast. NASA's operation here, the Wallops Flight Facility, hosts a number of missions, most of which are backbenchers to Kennedy Space Center. Wallops launches sounding rockets, weather balloons, and drones on missions that are important and yet somehow not sexy enough for mainstream media consumption.

But space launches are different. In 2008, the launch company Orbital Sciences Corporation changed the game entirely when it announced that its rocket, to be named Antares, would use Pad-0A here at Wallops Island. "The success of Orbital means the success of the spaceport," Bill Wrobel, director of operations at Wallops, tells me.

Orbital is a commercial space company with roots much deeper than Virgin Galactic or SpaceX. The company owns the Pegasus rocket, which drops from a jumbo jet and then roars toward space. They have been marketing it for launches since 1990, which makes it the first privately developed commercial launch vehicle in history. The company has also made a business out of refurbishing ICBMs from the nuclear missile inventory and using them for space launches.

Everything is looking up in 2013. Orbital is enjoying the opening of private space by the government when it lands a contract to resupply the International Space Station.

In April, Wallops launches a rocket carrying a dummy capsule in its shroud, the first time the spaceport launched anything that could reach the International Space Station. It's a real-world mission, with a paying government customer, and a promise of things to come. Kennedy Space Center is crowded with big players. Here, the vertical launch manifest is wide open.

Today is full of enthusiasm. "We're having four launches in four months," Nash tells the crowd. "We hope to have more."

But in a little over a year, this launch pad will be a hell of burning fuel and billowing toxic smoke. In the smoldering aftermath, the future of this spaceport suddenly will be in doubt.

ON JUNE 27, 1945, THE NATIONAL ADVISORY COMMITTEE FOR AERONAUTICS (NACA) and Navy rocketmen took over the beach at Wallops Island. They loaded a five-inch-diameter rocket with solid fuel and leaned on a spindly launch stand so that the tip angled toward the Atlantic. With a lick of flame and a high-pitched whoosh, the rocket rushed along its guide rail, leaping onto the air and arcing through the sky. Nearby, a NACA photographer captured the moment, the elongated bullet-shaped craft in midflight.

Just a few months before, in December, officials at the Langley Aeronautical Laboratory led by Engineer-in-Charge, John Crowley, organized a Special Flying Weapons Team. Its mandate: to create guided missiles and to secure the development of the weapons for NACA. World War II accelerated military weapons to the point that men of science could see warfare occurring at supersonic speeds. Which, of course, is exactly what happened.

The men who created this future formed the Auxiliary Flight Research Station, later named the Pilotless Aircraft Research Division. Its home was Wallops Island, at the time an obscure island on the virtually uninhabited coast of the lower Delmarva Peninsula of Virginia. The only things there were a Coast Guard station and some homes owned by sports fishermen. The engineers who worked early launches there enjoyed the release from stifling bureaucracy but had to endure primitive conditions and relentless insects.

The rocket on the beach that day was one of five that would tear through the air and slash into the ocean. Firing rockets was not something NACA knew much about, and the Navy ordnance specialists were there to prevent them from blowing themselves up.

The need to understand high-speed flight was urgent. The NACA gang knew that missiles and aircraft would be moving beyond the speed of sound, and engineers would need to design things that can steer that quickly. However, as one official NASA history of the era puts it: "Strange things happened in wind tunnels during tests at these speeds. Data readings, accurate above and below the transonic range, grew inaccurate within that range. A condition engineers referred to as 'choking' occurred when shock waves generated by air moving over a test model rebounded off the tunnel walls, interacting with the model. These frustrating difficulties led NACA researchers to consider new methods of obtaining test data."

There's no better way to know what any craft will do then putting it into the air and studying it. But that's not so easy with fast-moving missiles. This station would become a central location for such research, launching thousands of small rockets and tracking them with the latest radar and radio telemetry engineers could rig. And it all started when the first sounding rocket launched from Wallops Island on July 4, 1945.

As the years ticked past, the area became a magnet for innovative aerospace. "The focus of the Pilotless Aircraft Research Station expanded to include studying airplane designs at supersonic flight and gathering information on flight at hypersonic speeds," NASA documents say. "These tests included aircraft and missile designs from a variety of organizations and corporations including Douglas, McDonnell, Boeing, North American, Lockheed, and Grumman."

The place became so useful that it needed to become permanent. In 1949, according to NASA's history, "by use of condemnation proceedings, the government took possession of the island and later paid $93,238.71 in compensation to the previous owners." The facility tested components used for the Mercury and Apollo programs, and shot two monkeys inside rockets. (Both survived.)

In early 1961, Wallops became a real spaceport with the launch of Scout rockets, which deposited satellites in orbit. While the spaceport does not have a good path into geosynchronous orbit—for that destination, the closer to the equator the better—it offers the correct inclination to reach low Earth orbit. One day in the future, that would mean access to the International Space Station. But the Scout program ended in 1985, and everything at the Virginia spaceport stayed suborbital for decades.

In the 1970s, the facility took on the name Wallops Flight Center (later changed to Wallops Flight Facility) and started developing aircraft as flying science platforms. In 1981, Wallops became part of the NASA's Goddard Space Flight Center and was renamed again, this time to the Wallops Flight Facility. They inherited the least sexy program in the NASA portfolio, flying scientific balloons.

Balloons are cheap and easy tools to analyze the upper atmosphere and to test space hardware in severe environments. In the 1970s, NASA flew high-altitude balloons from remote locations across the planet—

from Palestine, Texas, to Brisbane, Australia. As payloads got larger, balloons crashed more often. This drop in reliability coincided with the consolidation of programs at Wallops Island, who witnessed the end of widespread balloon flights. From the thousands of flights conducted in the 1970s, balloon flights averaged fifty by the 1980s and twenty-seven by 1990. Having said that, the flights were longer, carried more valuable payloads, and had more reliable flight records. Wallops now sees more sounding rocket launches than balloon flights, but they continue to run a global long-duration balloon program from the Virginia coast.

Wallops became a trailblazer by involving the state government in launches. In 1997, the FAA issued the Virginia Commercial Space Flight Authority a spaceport license to operate at the NASA flight facility. In the state house, legislators lessened the insurance requirements and connived to get federal land handed over to build infrastructure.

Kennedy Space Center dwarfs Wallops in both size and attention. Leave aside the entire history of the manned flight program, if you can, and take it from a twenty-first-century spaceport-level analysis. The Cape enjoys infrastructure dividends from the Space Shuttle era such as long runways and huge buildings. KSC also neighbors very active Air Force launch sites and already provides for several prominent private space launch players, which only bolster their already advanced industrial base. The pads there have the infrastructure and toughness to withstand heavier rockets.

I ask Nash about all this in 2013. “We have less capability but we’re more flexible and more responsive,” he says smoothly.

It’s a good answer, since that responsiveness is on display behind him. This is a $280 million lunar mission—the Lunar Atmosphere and Dust Environment Explorer (LADEE)—and it is Wallops’ first deep-space launch. The spaceport authority upgraded Pad 0B for this moonshot by increasing the surface area with 4,600 square feet of reinforced concrete, extending the gantry height to accommodate the Minotaur V, and supplying five-hundred- and two-hundred-ton cranes that can hoist sections of the rocket during its on-pad assembly.

The seven-foot-long LADEE spacecraft inside the Minotaur V’s nose cone is small for its type, constructed in an attempt by NASA's Ames to create smaller, cheaper spacecraft that are modular. Wallops is a beneficiary of that attempt, as the seven-foot-tall spacecraft can fit in a relatively small rocket. Advances in power systems, propulsion, and communications could enable the creation of highly capable spacecraft that don't need massive rockets to loft them.

Smaller payloads are good news for Virginia's space industry. As cool as LADEE is to launch, the real steady work is launching to Earth's orbit. The ISS delivery contract lasts until 2015, but NASA will issue more contracts throughout the life of the ISS, at least until 2020. So there's a lot at stake.

The success of Orbital means the success of the spaceport. "I like to say all our wagons are tied together," says Wrobel ahead of the LADEE launch. "It's looking pretty good, if the missions continue to come."

YOU MEET ALL SORTS OF PEOPLE AT A LAUNCH. THE LADEE MEDIA center is overwhelmed by the attention, but I manage to corral a couple nervous scientists with experiments on board.

The centerpiece is called the Lunar Dust Experiment (LDX). It has a deceptively simple job: to collect and examine particles from the moon's exosphere, the name for a kind of super-thin atmosphere, about 1/100,000 the density of Earth's. These "surface-boundary exospheres" are the most common type of atmosphere in the solar system, or so I quickly learned.

"Yes, the moon has an atmosphere, it's just much more tenuous than ours," says Rick Elphic, LADEE project scientist at NASA Ames. "The dusty, flimsy mix of atoms and molecules in the lunar atmosphere is sure to have alien properties." Asteroids and other planets' moons have exospheres, so studying this one could be useful in other probes and even manned missions.

The science mission will last about a hundred days. During this time, LADEE will swoop to as close as 20 km from the surface of the moon, performing rocket burns every three to five days. The spacecraft has two spectrometers that will be analyzing the exosphere during these barnstorming maneuvers. According to Sarah Noble, LADEE's program scientist at NASA, the shallow angles are necessary. "We'll be looking sideways, through the atmosphere and dust, to understand what's happening just above the surface."

The LDX device will sample that atmosphere as LADEE moves through it at 1.6 km per second. Any dust in the detector will become a bloom of plasma that the LDX will record. The size of the dust is not as important as the presence of dust at all; it's not known for sure whether dust is present, which will make the instrument's first dust detection a reason to celebrate.

But there's another experiment on board, and I've been summoned into a trailer to meet the minds behind it, MIT Lincoln Laboratory fellow Don Boroson and NASA's Donald Cornwell. Their experiment is called the Lunar Laser Communications Demonstration (LLCD), which could mean the end of using radios to communicate with spacecraft. The testing begins before the spacecraft gets to the lunar orbit.

"We designed the latch (that covers the optical module) with a mirror on the inside," Cornwell says. "That way, we can test to see if the laser is working by bouncing it off the mirror."

When the spacecraft is in lunar orbit, Cornwell and Boroson will open the lid to the lasercomm space terminal by radioing a command to the spacecraft to heat up the paraffin wax holding the lid on. When it melts, the lid will pop open and the experiment can begin.

Because the wavelength of LLCD's lasers are 100,000 smaller than the S-band radio signals spacecraft currently use, a lot more information can be transferred using less power and smaller equipment. The equipment on LADEE uses half the weight and one-quarter the power of radio communications gear, Cornwall says.

Three devices on the Earth, located in New Mexico, Europe, and Massachusetts, will be available for data exchange. First the ground stations will track the spacecraft with a laser, and the LLCD then will use that connection to aim its own laser. When the first bits of information reach the ground stations, the researchers will then know they have done the first two-way laser communication with a spacecraft.

Today's spacecraft communicate with radio, but radiofrequency wavelengths are so long that they require large dishes to capture the signals. Laser wavelengths are 10,000 times shorter than radio, the upshot being that a spacecraft could deliver much more data than even the best modern radio system. For scale, the LADEE spacecraft that's carrying this laser experiment would take 639 hours to download an average-length HD movie using ordinary S-band communications. LLCD could download the film in less than eight minutes.

Cornwell and Boroson have been working on this experiment for years. They beat out other aspirants competing for room in the rocket, like some sort of sadistic reality television show. If the rocket explodes or the spacecraft doesn't respond, they will never get another chance to shoot lasers at the moon.

I ask where they will be during the launch. They tell me they will gather at a nearby seafood restaurant, rented for the evening by Orbital, that has an open bar and an unobstructed view of the launch pad.

The rocket is scheduled to take off that evening, and the cadre of press sits watching from the bleachers. Stephen Clark, the ace launch editor for Spaceflightnow.com is posting updates online, answering questions from less experienced reporters, and preparing his camera for the perfect launch shot—all pretty much at the same time.

At 11:27 p.m., the night turns into day as the Minotaur's engines ignite. Designed to deliver a nuclear warhead, the repurposed weapon is using its obscene power to expand the frontier of human knowledge. The white missile recedes into a pinprick of light.

A month from launch, LADEE will perform the most critical maneuver of its journey: firing its main engine for 197 seconds to slow the spacecraft enough that it is caught by the Moon's gravity. The craft must reduce its speed by 597 miles per hour or it is doomed to fly off into deep space. The flaming engine and position of the spacecraft will disrupt communication with the craft, leaving many anxious engineers and scientists. One of those will be NASA Ames Research Center director S. Pete Worden, whose agency designed and operates LADEE.

"It will be one of those heart-in-throat moments," he says.

When the moment comes, things go smoothly. As the craft enters its low orbit, LADEE's sensors show that the moon's atmosphere is dominated by argon, helium, and neon. "LADEE's dust instrument measured only a very tenuous dust cloud around the moon, far more tenuous than the apparent Pluto haze layers," writes Elphic. "The lunar dust cloud is caused by high-speed impacts of interplanetary particles on the moon's surface, not by haze formed from atmospheric constituents."

The experiment also discovers a strange cycle of freezing and release that may be repeated elsewhere in the universe. "As the moon rotates, the trapped night side gas molecules warm up at sunrise and are re-released into the atmosphere, and then some of them get re-trapped again the following month," Elphic says. "This cycling process is probably at work on Pluto as well, but instead of having our moon's twenty-nine-day cycle, it may be many, many years long."

On April 17, 2014, the researchers prepare for LADEE's death. "We'll burn all our fuel [to keep] from crashing into the moon," says Sarah Daugherty, NASA test director for the LADEE launch. When its fuel is exhausted, LADEE heads to its final dive toward the lunar surface. It gazes at the horizon on the way down, catching some new dust motes as the mission's Real Time Operations Team downloading data in the final minutes before impact.

What started on the beach in Virginia ends about five months later on the surface of the moon. When the cities and statues on Earth are gone, swallowed by the tectonics of our active planet, the remains of LADEE will be entombed on the lunar surface, preserved as a testament to mankind's relentless curiosity.

After LADEE, Orbital launches two successful supply runs to the ISS. The company is behind SpaceX, who is also delivering from Florida, but they're starting to develop a rhythm.

And then, October.

THE LIFTOFF LOOKS LIKE ANY OTHER, WITH A BLOOM OF EXHAUST and a roar as the Antares rocket starts to rise from the launch pad. The gantry leans crookedly as the exhaust washes over it, as it should. The rocket clears the launch pad's infrastructure, then pauses. Seconds later, the bottom of the rocket detonates.

The rest of the vehicle falls back to earth as uncontrollable flames paint the air. When the craft hits the pad, it explodes like a bomb. Ballistic arcs of flaming debris sail through the air. Video from the observer stand shows panic and flight worthy of a found-footage *Godzilla* remake.

Accident investigators will later agree that a turbopump failed, and whirling parts of the engine started tearing into the surrounding engine. With an engine destroyed, the rocket loses thrust and plummets. NASA and Orbital never will completely agree on the root cause of the turbopump's fatal collapse; it could have been foreign debris or a structurally unsound part. Orbital is partial to blaming a manufacturing flaw in the engine, which was made by Russia decades ago.

The launch pad, the only one configured to launch the Antares rocket, is another source of concern. When the smoke clears and the engineers survey the damage, they tally $16.2 million worth of repairs. Electrical systems are crisped, fuel plumbing ruined, and structures blasted and burned.

A little more than a year later, the Virginia Commercial Space Flight Authority reported that major repairs were complete. It seemed to be happy news—work crews fixed the site on time and, according to the VCSFA, within budget. But some, including the NASA inspector general, were left scratching their heads about how it came to pass that NASA paid so much to repair a launch pad it does not own.

What happened? To unravel the particulars, we need to go back to 1997, when the state of Virginia entered that Space Act Agreement to use land at NASA's Wallops Island facility. The idea was, and is, a sound one: have states run spaceports that can be rented to commercial interests, meaning Orbital.

But while the idea was clever, it seems the particulars of running a spaceport weren't entirely ironed out. From the start, there was confusion over who would pay for catastrophic damages. The chief counsel of NASA's Goddard Space Flight Center asked for some clarity at the time, and was told that the state agency would pay for insurance. But the insurance that the state obtained seemed to have loopholes.

NASA's inspector general reported that the agency "reviewed the policy in effect at the time of the Orbital mishap and found that, while it covers damage from aircraft and aviation operations, it explicitly excludes spacecraft and launch vehicles." So if an airplane had crashed into the Virginia launch pad, the insurance would cover it. But if a spacecraft damaged it, no dice. Essentially, then, NASA got insurance that satisfied the Federal Aviation Administration and met the letter of the law, but made little real-world sense.

In the aftermath of the October explosion, the Virginia Commercial Spaceflight Authority desperately needed cash to repair the pad and get back on track, so that the spaceport could continue to compete for ISS resupply missions.

Enter Congress. In December 2014, it passed the Consolidated and Further Continuing Appropriations Act. The bill includes one line item, written by four Virginia Congressional representatives and submitted through an appropriations committee, in which Congress directs NASA to reallocate $5 million from the agency's budget to fund repairs at Wallops.

Not too shabby on Virginia's part—your tenant breaks a launch pad, you get the feds to buy you a new one. But the VCSFA needed even more, according to the report, since the insurance they bought didn't cover the repairs. So they went to NASA, saying that the VCSFA "satisfied NASA insurance requirements" and that "the need to insure launch pads was not communicated by the Agency." Upon reviewing the back-and-forth between NASA and the Virginia spaceflight folks, the inspector general noted wryly that "it is not clear from correspondence between VCSFA and NASA that this issue was understood or agreed upon by both parties."

Eventually, NASA ponied up $5 million for Virginia's spaceport repairs. "According to NASA officials, the funding will come from other programs within the Space Operations budget, which includes the ISS and the Space Communications and Navigation Program," the IG report says.

NASA was backed in a corner—pressured by Congress and eager for the administration's commercial spaceflight effort to succeed. VCSFA did not comment publicly on the IG report. In a letter of response to the IG report, NASA said that while it did not have to pay for the repairs at Wallops, the agency "made the appropriate programmatic and policy-based decision" to give VCSFA the money. The agency response then goes on to promise that "NASA will continue to enforce existing requirements and procedures to ensure its commercial partners meet their responsibilities."

It's worth mentioning that Orbital has an FAA launch license that requires the company to cover damage to federal government property. The space company eventually paid $1 million for damages done to NASA's structures in Virginia. Asked for details, Orbital confirmed to me that they added launch coverage, but didn't disclose any more details because their "overall risk mitigation policy is a proprietary and competitively sensitive topic."

The debacle has changed the way the spaceport in Virginia handles insurance. When I ask the office of Virginia Gov. Terry McAuliffe, communications director Brian Coy told me via email: "In the wake of the disaster last fall the Governor and his team renegotiated a memorandum of understanding with Orbital that requires the company to carry insurance against future events. Previous to that (and to our administration) the spaceport was self-insured, which the Governor would tell you isn't the best idea when you're launching something into space."

As the commercial spaceflight industry matures and states continue to jockey for business, more complications like this will arise. There was no demand that a commercial entity insure its own property, and because NASA paid anyway, there is now little incentive for any other spaceport to do so if they think NASA will bail them out, too.

In his report on the incident, the inspector general sounded an alarm that, as of 2017, has not been heeded. "As NASA continues to rely on commercial companies, it is important to ensure all parties comply with procedures and clarify who pays for what in the event of a mishap."

In the case of Wallops, the taxpayer ultimately paid—at the expense of other space exploration efforts.

IN OCTOBER 2016, TWO YEARS AFTER THE LAUNCH PAD DISASTER, another ISS payload is ready for launch from Wallops Island. It's the same pad, the same destination, the same capsule, and even the same rocket. But this launch has to work. Orbital has been meeting the terms of its agreement with NASA by delivering cargo on an Atlas V rocket—relying on United Launch Alliance hardware to bail them and NASA out of trouble.

The rocket lifts off just before 8:00 p.m. The Cygnus capsule is packed with food and experiments for the ISS residents. In minutes the payload is in low earth orbit, maneuvering for a rendezvous with the ISS. Days later, it approaches cautiously and docks. The crew gratefully unloads the goods, earthbound scientists with their careers sunk into the experiments breathe a sigh of relief, and NASA officials and launch industry executives once again speak like humans instead of androids.

The astronauts then pack three thousand pounds of trash into the capsule. Cygnus decouples from the space station, aims itself at the planet, and reenters orbit in a fiery death dive. The spacecraft, garbage and all, burns up during re-entry into the atmosphere.

The mission is over. And the spaceport in Wallops Island has lived to see another day.

CHAPTER 6

ARMAGEDDON SPACEPORTS

THE STEADY THUMP OF A UH-1N IROQUOIS' ROTORS, A TELLTALE sound of wide helicopter blades cutting through the eddies of chopped air, can be heard for miles across the high plains of northern Montana.

No one in the military calls the aircraft by its Pentagon-given name—ever since it entered service in 1970, those flying and boarding the UH-1Ns call them "Hueys." They are tough, legendary birds, veterans of four decades of hard military service in the worst environments, from high-mountain rescues to humid jungle air assaults. Now in 2010, the Air Force employs them to provide overwatch for the personnel of Malmstrom Air Force Base, located just southwest of the city of Great Falls, Montana.

Col. Mohammed Khan, Jr., commander of the base's 341st Operations Group, pushes his head against the Huey's side window to get a look at the landscape streaking below. This section of Montana is a flat patchwork of tilled fields marked by the occasional unpaved road, trailer, or farmhouse. A wide band of asphalt, dotted with moving cars and trucks, is Interstate 15. The Huey intersects the highway and follows it at a brisk eight knots.

"There," Khan says over a headset, pointing to a rectangular chain-link fence surrounding a nondescript concrete slab. "You'd never know it was there, from the road, unless you know what they look like."

Below, a nuclear missile silo called India-7 sits less than a hundred feet from the interstate. There are no guard towers or high walls. It looks more like an unmanned weather station, rather than the cornerstone of national security. Dug into the ground beneath the concrete is a missile armed with at least one nuclear warhead. There are 450 silos scattered in Montana and Wyoming. Malmstrom has 150 of them.

America's only land-based ICBM is the venerable Minuteman III. These workhorses were to be replaced in 2020, but Congress and the Obama administration decided in 2007 to keep these weapons on alert until 2030, when they will be seventy years old. The Pentagon is spending billions on ICBM refits and silo renovations.

This is my first Huey ride, and my first trip to see the ICBMs up close. In 2010, when I first started examining the world of ICBMs, there wasn't much attention being paid to these missile silos. In the years after my 2010 trip, the media, politicians and the Pentagon itself will be drawn to these fields, after a swarm of scandals—from drug use in the ranks to drunken generals in Moscow—hits the nuclear missile force.

To me, the trip to Montana is a revelation. People don't consider ICBM fields to be spaceports. That's because they don't give much thought, not to mention appreciation, to the way the missiles work. Nuclear missiles carry pretty advanced spacecraft. Consider that the International Space Station orbits at 260 miles above the planet, while the Space Shuttle flew between 190 miles to 330 miles above sea level. The Minuteman III's payload of warheads has a maximum apogee of 750 miles. Anyone who says there are no weapons in space should take a look at the flight profile of an ICBM.

ONE CURIOUS THING ABOUT THE NUCLEAR ALERT FACILITIES, OPERated by the 341st Missile Wing, is the way they hide in plain sight. Long, one-story buildings, visible from highways, could be confused for residences if not for the microwave tower and a sturdy gate. The warning sign stating that use of deadly force is authorized to protect the grounds is another clue.

The building is actually a barracks called a Missile Alert Facility, where guards stay on five-day shifts to provide round-the-clock protection for the entrance to the capsule below. The missile control capsule sits sixty-five feet beneath this structure called a Launch Control Center (LCC). Captains and lieutenants and the occasional major spend twenty-four-hour shifts, or "pull alert," about eight times a month. Each capsule is hardwired with fist-size communication lines to monitor fifty missiles in remote silos, and is directly responsible for ten missiles in nearby silos.

The missiles under Malmstrom's responsibility are spread over 13,800 square miles of Montana landscape. The Air Force activated the first missile silo in 1963. When President John F. Kennedy responded to nuclear saber-rattling by Nikita Khrushchev with a warning that he had an "ace in the hole," he meant Malmstrom.

The Air Force offers to take me into an actual LCC for the *Popular Mechanics* article. It's one of those "through the looking glass" moments for me. Getting into that odd ranch house requires an elaborate routine, one that if done badly could lead to intervention by armed guards. I ask them to simulate the reaction of the guards if I were an interloper. A team of uniformed gunmen flanks me at gunpoint and throws me into the gravel, face first. Well, I asked for it.

And that's just the outside perimeter. Getting inside the launch facility requires another exchange a series of encrypted passwords with the security forces, filtered through handheld devices supplied to the Air Force by the National Security Agency (NSA). That enables us to board an elevator go to the capsule at the bottom of the hole.

The elevator door opens to a view of a four-and-a-half-foot-thick concrete-and-steel blast door, set in place in 1964. The eight-ton blast door is the only way in. Well, there's a second, last-ditch emergency exit, a sand-filled tube that leads to the surface. An escape would mean a long dig for the crew below, first through the sand and, since the tube ends before it hits the surface, several feet of dirt.

After a short elevator ride, we reach the foyer outside the capsule's blast door. The LCC is shaped like a pill and suspended by four pneumatic shock isolators capable of protecting the occupants and machinery from nearby nuclear blasts.

The massive door slowly swings open, and twenty-four-year-old Capt. Chad Dieterle, clinging to the metal handle, appears with a smile. He gives us the standard briefing before we enter the capsule, reminding us of the "no hands" policy inside. We're not to touch anything.

A freelance photographer and I, chaperoned by an Air Force escort, step inside one by one. As the last in, I close the door. I grip the metal handle and pull. Nothing happens at first, but then the door swings silently on its two bulky hinges.

"You'll want to go slowly and keep it from moving too fast," Dieterle says. "That's a lot of mass there." He glances at my feet, and I quickly pull my boot back as the door slides into place.

I scan the capsule. Banks of 1960s-era electronic cabinets line the wall. Very little has changed since the Kennedy administration, when this LCC was planted: Digital screens have replaced paper teletype machines, servers upstairs grant capsule crews internet access, and they now have Direct TV piped in for slow shifts.

But some equipment is awfully old. Dieterle pulls a floppy disk from a green console, part of the antiquated but functional Strategic Automated Command and Control System. Upgrading a nuclear missile launch facility or silo is disruptive—a very bad word for an enterprise that aims for 100 percent readiness—and the staff is not eager to replace anything that still works. "A lot of times older equipment is easier to use than anything new," Dieterle says.

Everything in the capsule has a backup. Commercial lines power the capsule, but generators above and batteries below can keep it operational during brief outages or global war. The air supply also has redundancies. The crew can make spare air with a hand-cranked device that separates oxygen from potassium superoxide. There is a graph in one of the many black-faced ring binders kept in the capsule that calculated the number of cranks it'll take to keep the crew alive, and for how long.

The pair in the LCC are responsible for making sure the launch facility and their five silos are secure, coordinating visits from maintenance crews repairing and modernizing the missiles. "We just don't sit around in our pajamas waiting to push a big red button," Dieterle says. Chimes fill the LCC, telling the crew of a change in status of the missiles, security systems or capsule. Sometimes the messages come in and the missiliers ask us to turn away. We don't look, even though it would be encoded jargon and gibberish to our untrained eyes.

Above all, the pair wait for the dreaded Emergency Action Message that could start a launch of the most devastating weapon mankind has ever devised.

IF THE LAUNCH CAPSULES ARE MISSION CONTROL, THE SILOS ARE THE ICBMs' launch pads. An Air Force major serves as our tour guide through the Little Belt Mountains to check them out in person. We drive around in a government van, past streams teeming with trout, farmlands stocked with healthy Black Angus, and woods thick with mule deer and pronghorns. The scenery is idyllic; at least it seems that way without snow and ice.

Even the Air Force personnel, equipped with detailed maps of the silos' locations, have a hard time finding the silos. My late afternoon visit to the Alpha-6 silo is marked by no less than three turnarounds on the two-lane highway that runs past Lewis and Clark National Forest.

We finally find the narrow, unpaved road that brings us to the silo. The air is still and quiet, broken only by the sound of a dog barking from one of the houses located less than a mile away.

For the cornerstone of America's deterrent posture, the nuclear missile silo is pretty humdrum—a flat slab of concrete surrounded by a chain-link fence. There are no guards. The silos here are dug into land leased from the property owners, often scant feet from highways, farmhouses, and pastures. One benefit of locating the silos on private lands is the protective mentality of the nearby landowners. Security officials make it a point to distribute phone numbers to them, in case of any suspicious activity. A call will bring an armed response.

The concrete slab is inset with a vaguely pentagonal block carved into the concrete, with two white rails leading from it. If the launch order is given, four explosive devices will drive a piston that flings the 110-ton concrete-and-steel door from the top of the ninety-foot silo. Everything is redundant—one charge alone could move the door. With all four going, the hydraulic force would fling the massive door through the security fence and a dozen feet beyond.

There is a pole with a video camera mounted at the tip, part of a 2010 security upgrade, but it wasn't even hooked up at the time. The launch facilities farthest away from the base get upgrades first, before the unreasonably harsh winter weather descends on Montana.

The perimeter is not as porous as it looked. Anyone going over the fence—be it a terrorist or a rabbit—will trigger a radar security sensor shaped like a long pointed stake. The radar at the tip paints the silo's surface with waves and alerts security teams in the Missile Alert Facility and inside the capsule if any new object appears inside the fence line. Every radar alert gets a response from Security Response Teams, two-man units who creep up on the site, fully armed, to investigate.

After you leave a silo base, your mind plays tricks on you. You start thinking in acronyms, and you think you see nuclear infrastructure everywhere you look. Every fence becomes a national security perimeter, and every farmhouse a nuclear launch facility. Every tractor-trailer is hauling weapons of mass destruction. It's well-informed paranoia in Great Falls. Every once in a while you'll be right.

IT'S JULY 19, 2013 IN CHEYENNE, WYOMING. LOCALS ARE CELEBRATING Frontier Days, an annual event tapping into "the spirit of the West." As part of it, Warren Air Force Base opens its chain-link gates and invites the public inside.

That's why 1st Lt. Lucas Rider, of the 90th Operations Support Squadron, is spending his Saturday chatting with rodeo fans that have come to gape at mockups of ICBMs. The airman, an Eagle Scout and former intern on *Your OWN Show: Oprah's Search for the Next TV Star*, puts a friendly face on Warren's secretive military operation. But his upbeat personality masks a deadly serious job: Twenty-four-year-old Rider is one of the airmen who await the order to launch a nuclear missile.

Rider is chatting up the locals when he spots his task force commander, Lt. Col. Erik Mulkey, approaching. He pulls Rider aside to ask, "Are you ready to head to California for two months?"

The young airman's response comes quickly: "I can be, sir." Two weeks later, Rider is on a plane heading southwest.

Rider has been chosen for a Glory Trip.

Three times a year, the Air Force randomly selects a Minuteman III ICBM from the missile fields, removes the nuclear ordnance, trucks the missile to Vandenberg Air Force Base in California, loads it with telemetry packages, and fires it at the Kwajalein Atoll, 4,700 miles away.

The missiles are aging, and anything that doesn't work right needs to be identified and fixed. Some airmen call these test flights the Super Bowl for ICBM crews.

The flights are big events at the spaceport, as well. Vandenberg Air Force Base, seventy-one miles north of Santa Barbara, doesn't bathe in the same the historic glow generated by the Apollo and Shuttle eras of spaceflight. If they know it at all, people associate the base with nuclear weapons picketers who sometimes show up at the gate. Martin Sheen was once arrested at the front gate, although his public affairs people tried to broker a peace via his son Emilio Estevez, who owns a nearby winery.

In September 2013, I trek to the base to cover two of these launches for articles in *Popular Mechanics*, observing the preparation and launch of a pair of ICBMs for a magazine feature. On the website, we call it "a dress rehearsal for Armageddon."

The Glory Trips are designed to test the missiles, but they have the added benefit of boosting the morale of the missile crews and maintainers who keep watch on America's nuclear arsenal. Rider, however, does not come off as demoralized. The young airman tells me that he hadn't even heard of ICBMs before joining the Air Force. Now he is one of the men with his hands on the switches of the most powerful weapons ever created.

I meet Rider and the other sincere but amiable twenty-something Air Force ICBM launch crews, who came in from Warren and Minot Air Force Bases. The Minuteman III to be fired in GT-209 travelled by truck from Wyoming to California, just like the launch crews.

The GT-209 mission patch, which everyone here wears on their left arm, is a grey isosceles triangle depicting a snarling, three-headed hound. The mutt is Cerberus, three-headed dog that guards the gates of hell. It's a not-so-subtle nod to the test of three dummy warheads, each representing a nuclear warhead capable of destroying a good-sized city.

The launch is scheduled for September 21. ICBM test launches happen in the early morning, US time, so they disrupt as few people's lives as possible. Airplane traffic has to be diverted on the West Coast and Hawaii. Boat traffic on the California coast shuts down. Even the local train line, built sixty years before Vandenberg Air Force Base was established, stops. It's like a large part of the world pauses for liftoff.

A launch countdown is really just a series of smaller countdowns. Inside the ICBM Launch Support Center (ILSC) at Vandenberg, uniformed members of the 576th Flight Test Squadron and civilian contractors sit at their consoles, reading data flashing on the screens. The airmen wear green flight suits, while the civilians dress business casual in jeans and slacks. Most of the civilians wear ties. All speak jargon into headsets and consult exhaustive checklists about battery voltages, pressure readouts, telemetry feeds, you name it.

The ILSC fits about fifteen people. Two large, black-and-white video screens dominate the front of the room. One shows the silo door at LF-10; the other alternately displays weather radar images and video feeds from Kwajalein, the watery impact zone for the dummy warheads. In between the screens are digital clocks that display countdowns and a red-yellow-green bar of lights indicating the launch range status.

Reams of data unspool from a Microline 184 dot-matrix printer.

Every so often, one of the civilians folds the paper and places it on an ever-growing pile. The out-of-date printer—like the Minuteman III and its launch infrastructure—was designed in the 1960s. If it works, don't replace it.

The launch director, Capt. Denise Michaels, is in the middle of this scrum. She serves as the authority in the room and the point of contact between ILSC and the ultimate head of the missile-shot: Col. Lance Kawane, commander of the 576th flight test squadron.

Lt. Jim Gutierrez, the launch's countdown compliance officer, is seated next to Michaels. His voice is the most common one heard from the ILSC. At T-minus 58, his official launch countdown starts. "Prepare to initiate," he says in a ready-for-drive time radio voice. One by one, everyone checks in. A moment later, off mic, he makes a fist and simulates rolling dice at a craps table: "C'mon baby, no holds!"

At T-minus 5 minutes, the countdown pauses for another mandatory check of the people and hardware, spread across half the globe. One by one, everyone reports systems are "go." In the ILSC, Michaels calls Kawane (who was in another control room, in another building) for permission to restart the countdown. It's given, and the range condition light switches from yellow to green.

Rider and 1st Lt. Nathan Larson of the 321st Missile Squadron take their places at a console. Larson and Rider fix their eyes on the digital clock, hands on the switches. With sixty seconds left until launch, they simultaneously turn their hands and began "terminal countdown." Seconds later the launch control center rumbles. Just a few hundred yards away, the ICBM is roaring to life in its silo. Four ballistic gas generators blow the 110-ton silo door open.

In the ILSC, the black and white video feed shows the silo door sliding away. I describe it in the magazine as happening "quicker than the camera's frame rate can handle, and the image skips like the lid of a vampire's coffin opening in a silent movie." The silo's exposed dark hole belches a white-hot jet of flame and smoke. The missile leaps from its hole, cutting through the roiling gasses and trailing a flickering lick of orange flame.

Now, the entire Glory Trip rests with one guy. Everyone at Vandenberg calls Mission Flight Control Officers by their acronym, MFCOs (pronounced "Miff-Coes"). They have an unpopular job: to destroy the rocket if it veers off course. The MFCOs motto—"Track 'Em or Crack 'Em!"—puts people on edge.

Lt. Dylan Caudill, twenty-three years old, is in the MFCO hot seat. Caudill's screen displays a digital map with lines that the missile cannot cross. A GPS transmitter tracks the missile's location, an X marking the spot on Caudill's terminal. An elongated white oval extends in front of the X—this is the projected debris track in case the missile crashes.

If the Minuteman III crosses any safety lines on his map, Caudill will activate the switch that detonates explosives along the seam of the rocket. This is called "sending a function." It would not cause an explosion. Rather, it would crack the missile so the fuel rushes out before it ignites. The ICBM would then spiral harmlessly into the ocean—and the MFCO would have to answer for why.

"I've never sent function to a vehicle," Caudill says. "I almost did, once. It was terrifying."

After a few seconds of vertical flight, the Minuteman III tilts and begins its journey west. Data about the missile's yaw, pitch, and pressure readings streams into Caudill's console. "Because the Minuteman III pitches and rolls, sometimes even during a nominal launch, these screens can get spiky," he says. The High Accuracy Instrumentation Radar at Vandenberg follows the missile's progress as it continues its steep ascent.

After a minute of flight, the Minuteman III is at an altitude of 100,000 feet, but just eighteen nautical miles away. A group of airmen at a nearby viewing site sees the missile as a diminishing orb of light over the Pacific. Suddenly, the light splits. A second light blinks as it eases away from the first. This is first-stage separation, when the spent tube falls away from the rising rocket and tumbles back to the Pacific.

If you can forget that they are weapons of mass destruction, they can be quite beautiful.

Once in space, the missile's nose opens and exposes the three dummy warheads. Mini-thrusters on the missile's cone fire to steer the shroud away from the bulk of the rocket.

The second stage separates again at 126 seconds, when it is at 300,000 feet high and 120 nautical miles away. The second stage motor fires, producing 34,000 pounds of thrust to push the missile higher. Tens of thousands of pounds of solid fuel combust in a single minute.

The once-towering rocket has become a stub, less than half the size of what left the ground. At 180 seconds after launch, the third stage falls away. Only the section carrying the warheads, the post-boost stage ("bus"), remains as the missile reaches its astronomical apex and begins its descent.

During the plunge back to the planet, the bus's liquid-fuel propulsion system rocket engine (PSRE, or "pizz-ree") fires to adjust the trajectory of the warheads. Without a PSRE, a nuclear missile can only hit one target per missile. The system can also fire decoy warheads, confounding radar and defensive systems.

The PSRE fires its rockets to slow the spacecraft as it releases the warheads. A mechanism called a zero impulse bolt releases the warheads without changing their ballistic trajectory, leaving them locked on target.

IT'S A QUIET NIGHT ON KWAJALEIN ISLAND. IT'S ALWAYS QUIET HERE. The atoll, which the military calls "Kwaj," lies in the deep Pacific, 2,400 miles southwest of Honolulu. Only a thousand people live on the 6.6-square-mile island chain, and nearly all of them work at military installations. On the evening of September 21, 2013, most of those people are watching video screens, radar returns, or streams of telemetry data.

GT-209's three dummy nuclear warheads are hundreds of miles away but closing fast. Each of six-foot-long triangular wedges blazes a trail of glowing plasma formed by the air friction of hypersonic flight.

Out in the Pacific, the warheads appear as bright pinpricks—shooting stars following each other to the horizon. Radar in domes follows the warheads' progress. When the mock W87 warheads approach, batteries of high-definition, long-range cameras capture their flight for analysis.

For such long range shots, and without guidance for most of the flight, ICBM warhead accuracy is impressive—the shot's circle error probability is reportedly as little as five hundred feet. To hit Moscow, ICBMs would aim at a point 186 miles east of the city to take into account the rotation of the planet during flight.

At Vandenberg, all eyes are fixed on black and white screens showing the feed from Kwaj. A glowing dot appears and blooms into massive flare on the screen. The orb is followed by two others. When the warheads splash down, an underwater impact location system registers their locations by the sound of their impact on the surface.

In the ILSC, the lights bloom and wink out of existence on the screens. The applause begins, high fives exchanged and headsets worn for hours put down on consoles.

The room drains of personnel. Most of the civilians are heading home, but the airmen have to produce a "quick-look" report on the basics of the launch. In the Pacific, the warheads settle to the bottom of the ocean, too deep for recovery. GT 209 is over. Everyone will take off the Cerberus patches from their uniforms.

Mutually assured destruction has again been confirmed.

IN 2015, THE PENTAGON, AFTER ABOUT TEN YEARS OF STUDYING THE issue, finally decide on a direction for the replacement of the Minuteman III missiles and launch facilities. They decide to keep what's in place.

The news comes quietly. The Air Force releases a document called a "request for information" to industry. An RFI basically says to defense contractors, "This is the direction we want to go, now tell us how you'd make it happen." This RFI makes it clear that the Air Force is planning to replace the Minuteman III ICBM and replace it with a new launch rocket. But these new systems will launch in the same silos where they sit now.

Here's how the document puts it: "The Government is preparing to acquire a replacement for the MM III intercontinental ballistic missile system that replaces the entire flight system, retains the silo basing mode while recapitalizing the infrastructure, and implements a new Weapon System Command and Control (WSC2) system."

One earlier RAND report described the current silo set up as "vulnerable to high-end enemy." But it also dubbed it "cheap and familiar." No wonder it's the leading plan within the Pentagon.[6]

Once upon a time, the Air Force looked into game-changing strategies. One Air Force official in 2010 told me it was time to "dust off some older ideas." Intrigued, I took a look at some of those concepts. In the 1980s, the Pentagon considered more than thirty other ways to keep nuclear missiles on permanent alert. Those old plans have plenty of creativity, maybe too much.

The most un-PC one is the Land-Mobile Midgetman missile. These stunted missiles would be housed inside blast-hardened vehicles that can take ten times the overpressure as a tank. Popular in Russia, the dense population and increased security risk of mobile nukes make this problematic in America.

Nukes buried in shallow water or sunk deeper, as studied under the Pentagon's Orca and Hydra programs, give them great security and survivability. But treaties ban stationing them farther than twelve miles from shore, making them vulnerable to attack.

Digging missiles into mountains is also problematic, for its expense, but offers certain advantages. The missiles would be buried thousands of feet deep, and transported by rail through an extensive maze of rock tunnels. In a second strike scenario, equipment could be installed to dig tunnels or shafts to the surface—so the retaliatory missiles could be launched. Not many Congressional districts would lobby for this base.

Then there's the rail garrison plan. The launchers loiter in train

6. A new missile does not mean a new warhead. "The new weapon system will use the existing Mk12A and Mk21 Reentry Vehicles (RV) in the single and multiple RV configurations," the RFI said.

alert shelters. During an alert, the trains would roam over a large rail network becoming elusive targets. Final design for project was completed in 1990 when this concept was also canceled. But the plans are ready to be dusted off whenever there is political will for it. There is certainly some "not in my backyard" pushback expected from this plan, and the US rail system, clogged with merchandise and aging from a lack of maintenance, is not as robust as we'd need for this plan to work within a reasonable cost.

The safest place for nukes might be off the planet. Positioning nuclear weapons in a space station is a pretty clever idea, so much so that a 1969 treaty was signed to prevent it from happening. Besides being expensive to launch and maintain, space junk and anti-sat weapons put such space silos at risk.

But parts of the concept may still be alive in Russia. One of the benefits of nuclear weapons in space is that they can strike from unexpected directions, especially across the poles. That was true during the Cold War, and it is true now. Back then, the Soviet Union was concerned about the short range nukes pointed at its homeland, since a first strike could take away their ability to respond. A nuclear warhead that could reach orbit on a large rocket could orbit the earth a few times before dropping on an unsuspecting target. "Given global missiles, the warning system in general has lost its importance," Nikita Khrushchev remarked in 1962, sending diplomats and national security analysts at the time into a frenzy—and toward that 1969 arms treaty.

Fast-forward to 2014. This time, it was the former head of a Russian government research institute, Maj. Gen. Vladimir Vasilenko, talking about ICBMs large enough to put a nuke in orbit and land it in the continental United States. The Russians were looking for ways to beat the new, ground-based missile defense systems based in Alaska. "The huge payload potential of the heavy ICBM makes it possible to equip it with a variety of means of missile defense suppression," he added.

What he's describing, then, is not a warhead but a spacecraft with the suicidal intention to deliver nuclear warheads and decoys to a target from space. The remarks stoked fears that have become realized with the unveiling of the RS-28 Sarmat ICBM in 2016, possibly to be deployed as early as 2018. While the Russians haven't released specifications, the sheer size of this rocket—more than 100 tons—makes is a likely candidate for this orbital bombardment style of attack.

The United States has no such heavy ICBM under construction or even planned. It's not a politically or economically feasible option.

So for Global Strike Command, it's back to a widely distributed network of silos. Given the importance of nuclear weapons, it would make sense to dig fresh holes. But the legal and political fights associated with such a base are hard to fathom. Environmental reviews, land-rights agreements, and security breaches during construction would bedevil the effort from day one. That leaves the United States reliant on the literally leaky infrastructure that already exists—what has been underground since John F. Kennedy was President.

The RFI does hint that a change in the way the missiles are launched could be in the offering as part of the new weapons control system. "The government is also exploring options to reduce/streamline the current LCC/LF architecture," the RFI document says.

So fewer launch control centers—and therefore fewer airmen—would be in control of more missiles. It will be interesting to see how the mix of automation impacts reliability and security. Will people feel better or worse with a robotic system with a digital finger on that figurative red button?

ON AUGUST 2, 2016, A BLACK TRUCK PULLS UP TO VANDENBERG AIR Force Base. It is another payload delivery, but this is no nuclear missile or dummy warhead. In the bed of the truck are three climate-controlled boxes, carrying twin satellites and assorted gear to launch them.

The satellites themselves are fairly revolutionary. They are destined to be part of Iridium NEXT constellation. In late July, before they shipped, Iridium's chief executive officer Matt Desch gushed over the 1,900-pound objects. "After more than seven years of effort, the first of our next-generation satellites are finally ready for space," he said. "This program replaces the largest commercial satellite constellation in space with state-of-the-art technology and new capabilities."

The idea is to try setting up this network in low Earth orbit, a scheme that has failed before. But improvements in tracking antennas, small satellite design, and solar cells in spacecraft arrays are prompting Iridium to give it another shot.

For Vandenberg, hosting a private launch is welcome. These days, government spaceports need to be open for business. This launch is a window into the base's less-lethal job as a commercial space launch center.

ICBMs aside, Vandenberg is important as a spaceport. It's always been a workhorse, a home for launches that don't fit on the launch manifest in Florida. The United Launch Alliance lofts spy satellites and other national security payloads from here. These are often National Reconnaissance Office and military communications satellites, which are placed in polar orbits with global coverage. Flying south, satellites can get there from California easier than leaving from Florida's flight paths, which fly east to get over open water and away from the populated coast.

Vandenberg is enjoying its role as a host for commercial space. The trailblazer, as usual, is SpaceX. The private space company wouldn't have gotten off the ground if Vandenberg hadn't accommodated them.

In 2004 Musk's firm was renovating Vandenberg's SLC-3W in preparation for the maiden flight of the Falcon 1. "Although we were able to start with the existing concrete foundation of an old Atlas II pad, there was literally nothing else in place," Musk wrote on a company blog. "Compared to other launch pads at Vandenberg, we don't require much, but there is still a lot of work to do to have a professional launch infrastructure."

That included installing new plumbing to connect lines of liquid oxygen, kerosene, helium, and nitrogen. All of these umbilical lines need to be designed to detach (quick-detach umbilicals, Musk calls it) as the rocket lifts off. "Right now, all the electrical and communications wiring is in process of being installed and connected to the base grid," he wrote. "The water deluge for heat and noise suppression on launch will just use standard base water pressure, as Falcon 1 doesn't need anything more to meet specifications."

Maybe so, but there is more to spaceports than just specs. SpaceX was about to bump up against the neighbors. "Lockheed, our next door neighbor at SLC-3E, is preparing their Atlas V pad for launch," Musk noted in July 2004. "They have been good, courteous neighbors and hopefully Lockheed sees us the same way."

They didn't. SpaceX had been launching its Falcon 1 at Kwajalein, an inconvenient but safe place to try this sort of thing. They crashed three rockets, but were still begging to be allowed to launch in California.

By 2006, there is plenty of spaceport drama at the California airbase. Lockheed Martin, on the launch pad next door, convinced the base to evict SpaceX, even after they had completed the renovations and were getting close to test-fires. The official reason was "safety concerns." Of course, Musk's upstart company's stated reason to exist is to steal launches from the existing players. "I had certainly come to the conclusion that if something didn't happen to improve rocket technology we would be stuck on earth forever," he said in 2016 of the company's origin. "The big aerospace companies had no interest in radical innovation. All they wanted to do was make their old technology slightly better every year, and sometimes it would actually get worse."

SpaceX finally succeeded in launching a Falcon 1 from Kwajalein in 2008. In 2010, they moved operations from the remote Pacific Island to Cape Canaveral. They also graduated to the Falcon 9—the number designating the number of engines on board this more powerful rocket.

But SpaceX wasn't through with Vandenberg. SpaceX has been launching from Vandenburg's SLC-4 East since 2012. The company also began a five-year lease of SLC-4 West in February 2015. The goal, as always, is reuse, and LC-4W will become a landing pad for empty first-stage boosters of the reusable Falcon 9 and Falcon Heavy launch vehicles.

I can't help but think of the young Air Force officers working at the base. These crops of MFCOs, engineers, mission control operators, the launch pad workers, and radar operators are at the forefront of the new space industry.

A young officer like Lt. Caudill has a resume of a 1950s sci-fi hero. His office is a spaceport, supporting two space launch companies with different rockets on the pad. From mission control, he's watched ICBMs loft into the air and sent the heaviest rockets in the US launch secret payloads into orbit. He's worked the room when the unmanned X-37 spaceplane landed, skidding to a halt in between its months of secret missions in orbit.

Caudill says he knows he's a lucky airman. The future, whatever it looks like, will belong to spaceport range rats like him.

CHAPTER 7
THE ULA/ SPACEX FEUD

On launch night, the Cocoa Beach Pier charges admission. Throngs of people fork over ten dollars to stand on the fishing pier, where they can get an unobstructed view of Kennedy Space Center. Most gawkers come from the cluster of bars and restaurants at the pier's base and mill around with beers or plastic cups of tropical drinks in hand. It's a beach vacation vibe with a sci-fi twist.

Experienced locals gaze north for any spark of light from Space Launch Complex 37. There, a Delta IV is ready to launch. The payload this evening—February 20, 2014—is a GPS satellite owned and operated by the US Air Force but used by the entire world.

The light of ignition flares bright enough to illuminate the clusters of people in folding chairs and towels on the beach. Their faces are uniformly oriented toward the horizon like satellite dishes or sunflowers. Within two minutes, the rocket is just another dot of light in the starry sky. By late the next morning, telemetry reports confirm that the sat has reached its 11,000-mile-high orbit. It's another success for Cape Canaveral's go-to launch providers, the United Launch Alliance (ULA).

The Florida Space Coast didn't go dormant after the Space Shuttle's retirement. It did suffer nearly 10,000 layoffs, a steep reduction that reinforced the idea that Cape Canaveral and its surroundings are fit only for celebrating the past. But Kennedy Space Center is still "heavy-lift country," where the Uncle Sam turns when he needs a critical piece of hardware lofted into orbit.

Kennedy Space Center in 2014 is not the same spaceport it was in 2002, when the first Delta IV launched from there, or in 2011, when the shuttle retired. There are big changes going on behind the gates. New companies are setting up shop. The state of Florida is snapping up and rehabbing facilities, in hopes they can rent them to private firms. The state is also seeking to build a new spaceport adjacent to the federal facilities. Legacy tenants like ULA are trying their damnedest to keep their seat at the table, pushing into new hardware and missions.

The tears of 2011 are dry. KSC is not dead. The historic spaceport is becoming something else for America—a roost for new companies with dreams of reaching orbit. Those buildings that emptied after the shuttle retired are refilling as America's premier spaceport carves a place for itself in the twenty-first century. But the process can be messy and, sometimes, even explosive.

FRANK DIBELLO HAS ONE MAIN WORRY: WATCHING HISTORY PASS BY Cape Canaveral. At first glance, he doesn't appear or sound like much of a visionary. As the CEO of Space Florida, the agency created by the state to promote spaceflight, he speaks and dresses like a businessman. No one would confuse his short, stout frame for that of a retired fighter pilot or astronaut. But when you get him going, DiBello is as forceful of an advocate for private spaceflight as exists.

I first meet DiBello in 2014, on his turf—an office in Exploration Park, the tech park that's home to space-related companies that do business with the spaceport. He has a soft hand but a firm handshake. DiBello may look like an accountant, but he's an agent of change who wants to knock NASA and state politicians out of their lethargy. "They keep hoping the industry of twenty years ago will come back twenty years from now," he says. "But the past doesn't ensure the future."

Spaceports work together to boost the industry as a whole, but they are also involved in knife fight competitions over contracts and tenants. The jovial man scowls when you ask him about SpaceX's 2014 choice to build a spaceport in Texas. "I am mad as hell that we could not offer him a comparable alternative business site and environment here in time," says DiBello. "Something has to change."

His business chops always mixed with future tech. DiBello co-founded Aerospace Capital Partners, an infrastructure investment fund focusing on aerospace, technology, and telecom initiatives. DiBello also founded KPMG Peat Marwick's Commercial Space and Advanced Technologies Practice and managed the firm's Aerospace Industry Practice for more than fifteen years. "He comes from a world of finance and investors," says Dale Ketchum, Space Florida's chief of strategic alliances. "He speaks capitalism very well."

In 2009, the state of Florida tapped DiBello to drag the Space Coast into the commercial world. He tells me that their "frozen-in-time attitude won't get the job done in an era of rising competition abroad and within the United States."

DiBello says the current crop of players have an approach better suited to the high-risk, high-cost world of spaceflight than their predecessors. "Before, it was hardware companies seeking to become information companies in space," he says. "Now, it's the opposite. IT companies are seeking to become space companies. Paul Allen, Jeff Bezos, and Richard Branson see a future in the next frontier. These people understand markets."

What makes DiBello a great interview is not his business acumen, but his zeal as a futurist. A conversation with him about real estate and state politics drifts into the possibilities of spaceflight on those below. Spaceplanes that can land at a network of runways across the world will enable a new wave of organ donations. Small satellites, and the rockets that launch them, will open up new markets in streaming information to and from electronics embedded in clothes. Point-to-point travel will bring conventioneers from Dubai to Orlando in two hours. Microgravity experiments will revolutionize food production, and open the new field of space agriculture. Everything is possible, everything is pragmatic, and at the end of the day someone always makes money.

DiBello's moves set the stage for the spaceport businesses that support the actual launches. "Florida has to be more than just a launch state," he says.

To see an example, I head over to the Space Life Sciences Lab, formerly run by NASA. This is where scientists and techs prepared space experiments for the Space Shuttle and the International Space Station. I ask about one of my favorites, an experiment in the shuttle to grow sunflower seeds in space. The purpose: to see if the signature corkscrew motion at the root's tip depended on gravity. It doesn't.[7] The facility's operations manager, Carol Ann Taylor, leads me to a climate-controlled vault where flight-ready sunflower seeds and their understudies were prepped, each at a specific stage of life.

Any space experiment comes with a logistical demand, and usually that means a company, university, or government agency needs a lab close to the spaceport to get it done. These labs are not just empty spaces, but demand backup power, special venting, clean rooms, and large cargo bays for special deliveries. DiBello anticipated this demand and in 2010 Space Florida took over the Life Sciences Lab.

These days it seems like every room and floor houses some experiment. The facility also hosts thirteen tenants, some working on space-related research (including a few destined for the ISS), but most others just in need of a place to pursue their earth-bound experiments.

This lab could have a vibrant future as more companies and colleges see that trips to space are possible. "People have to see that this can be done outside of NASA," says Taylor.

AT CAPE CANAVERAL, THE FUTURE IS BEING BUILT ON THE LEGACY of old launch pads. Howard Biegler, the human-launch-services lead at ULA, walks across Launch Complex 41, which has endured blastoffs of dozens of Titan and Atlas launches. Satellites, solar probes,

7. I learned about this obscure experiment while researching a previous book, *Sunflowers: The Secret History*.

and Mars-bound spacecraft have left Earth from this spot. "It makes sense to reuse the old Titan pads," he says. "They cost billions of dollars in infrastructure to create, in reinforced concrete and steel."

There are fresh red Xs painted on the weathered concrete, spaced twenty feet across. These are the positions where the legs of a new support tower will be placed. The top of the tower will feature a walkway for astronauts to board spacecraft, something that the Cape has not seen in three years, since the end of the shuttle in 2011.

Asked who painted the Xs, Biegler smiles and says proudly, "I did!"

Human spaceflight, as ULA officials will tell you, does not pay the bills at Cape Canaveral. The good money today is made on Pentagon launches. The National Reconnaissance Office, NASA, and the US Air Force are some of the ULA's steady customers. Between 2011 and 2015 the alliance launched eight to ten rockets each year, double the amount when the shuttle was flying.

But lofting astronauts is a source of pride and excitement, and ULA and Biegler are not immune to this. The rehab of this SLC-41 is a symptom of this newfound attraction to manned flight.

When I visit in 2014, two spacecraft are to be flown from SLS-41: the Dream Chaser spaceplane and the Boeing CST-100, called the Starliner.

The Dream Chaser would launch on an Atlas rocket, detach to reach space, and coast for a runway landing like the Space Shuttle. "The way I characterize it is the Space Shuttle is sort of like the big moving truck that takes you from New York to Florida," Mark Sirangelo, corporate vice president and head of SNC's Space Systems, told me in 2014. "It has a cab up front and a huge back to take all the big pieces and all your furniture and things. We're sort of the SUV for space."

Several entities have hitched their wagons to this spaceplane, from the spaceport cheerleaders in Huntsville, Alabama, to the United Nations Office for Outer Space Affairs in Vienna, Austria. It has never flown, but anticipates a launch in 2019.

Sirangelo gives one of the best pep talks about point-to-point spaceplane travel I've ever heard. "We're planning to bring the vehicle back to different places around the country," he says. "Because we have no hazardous material on board, our engines and our systems are non-toxic, we can actually land in any airport that can take a 737. And we can fly home inside a cargo plane. Unlike the shuttle, we don't need a special carrier. It's actually movable by truck if we wanted."

The promise, as with all spaceports, is to spread the influence of spaceflight around to places and people who have never experienced it firsthand. In terms of education and economy, it's what every municipality wants for the twenty-first century.

"The idea here—and obviously we have to work through a lot of different permissions—is, wouldn't it be terrific if we brought America's space program to America?" Sirangelo says. "And instead of having people come to see the shuttle land in Florida or California, we could plan landings on our return trips to different places around the country where those states or the universities or the high schools could bring the students out and get to see the space program."

In 2014, Dream Chaser is vying for a NASA contract to ferry astronauts to the International Space Station. It stands a good chance, too—except later that year SpaceX and Boeing will edge the private space firm out after surviving several rounds of funding. But Sierra Nevada garners enough government funding for a test flight, kindling hope that the design will prove itself and be adopted by the private space world—in the United States or elsewhere.

However, the support tower at SLS-41 is to be used to support Boeing's Starliner capsule. NASA selected the CST-100 and awarded Boeing with $4.2 billion to make a spacecraft that could reach the ISS, and maybe later be used to dock with inflatable space habitats made by Bigelow Aerospace. The Starliner spacecraft is expected to fly unmanned in June 2018, with a possible first crewed test flight in late 2018. When it does launch, ULA will be providing the ride and it will leave from SLC-41.

If any company can sit back and enjoy the space status quo, it's the United Launch Alliance. They are entrenched at the Cape, good at what they do, politically savvy, and proven to be reliable. But the winds of change are sweeping around this monopolistic launch provider, and the spaceport is changing to keep up.

It's easy not to like ULA. The firm, formed in 2006, is a joint venture between aerospace giants Lockheed Martin and Boeing. ULA was formed to put an end to bitter competition between the aerospace giants. The two were going at each other like gladiators in the 1990s and 2000s, fighting for the hundreds of billions at stake in national security launches. The fight got ugly; accusations of industrial espionage were leveled and federal law enforcement was summoned. In 2003, federal authorities charged two former managers at Boeing with stealing secrets from Lockheed.

The battle hurt national security—someone had to launch the damn satellites. The solution was to have the companies form a fifty-fifty partnership. The Federal Trade Commission (FTC) gave its antitrust clearance in October 2006, but with caveats aplenty, saying the formation of a joint venture "is likely to cause significant anticompetitive harm." But this is space launch. The most important criterion is not fairness or frugality, but reliability.

At the time, private space was limping along and Russians and Europeans had locked down the commercial satellite launch market. The national security benefits were enough that the FTC judged the venture worthwhile.

The ULA has been trying to drive down prices, and it's also looking at expanding into other markets. The alliance doesn't chase commercial satellite launches, a market dominated by the EU, China, and Russia. The organization has performed more than ninety successful launches since 2006, an enviable record. Less happily: ULA's Evolved Expendable Launch Vehicle program exceeded its original per unit cost by 286 percent.

Over time, the monopoly structure allowed for ways to trim those costs. For example, the Air Force began to make block purchases, which makes launches cheaper by stabilizing demand. "This disciplined approach saved the government and taxpayers approximately $4 billion while keeping our nation's assured access to deliver critical national security assets safely to space," ULA argues.

The plan—warts and all—is working. When the Space Shuttle retired in 2011, ULA was still lofting satellites into space with regularity. America cannot launch astronauts but, thanks to ULA, spy satellites, Mars missions, and deep space probes leaves from US soil. It is the backbone of US spaceflight.

Yes, it's trendy to hate on ULA. And they are not easy to defend, since they cost a lot of money and, until now, have had no competition. But despite having cornered the market on government launches for years, ULA's job is not easy. Not many entities can launch payloads for a fickle military and NASA customers as regularly and successfully as ULA has done. The space launch landscape is now shifting under ULA's feet, leaving it with full manifests but an uncertain future.

Customers accept high costs. ULA launches might be expensive, people thought, and the Air Force and NASA's own rules probably drove up the cost even more. But whether the rockets rise in Russia, French Guiana, China, or the USA, launches are expensive. It costs money to assure the precious payloads get into space safely.

At least that's the way space launch used to be regarded, until SpaceX entered the market.

ELON MUSK IS LATE TO HIS OWN PARTY. WELL, IT'S NOT HIS PARTY, but he's in New York City to receive *Popular Mechanics*'s 2012 Breakthrough Award, an annual ritual for the publication. The recognition comes with a cover story and a ceremony, held at the Hearst Tower on 57th street.

I'm waiting for Musk to show, sweating under my suit collar. I'm the one who told him when to show up, but the time I told him was an hour past when he's actually due. I stand guard at the building entrance, hoping the billionaire will arrive soon.

Just the year before, I lobbied in magazine meetings against naming Musk as a recipient. But actions speak louder than words and in 2012, he launched successful missions to the International Space Station and broke ULA's monopoly on government launches by winning two contracts.

The launch provider is also selling its services to the private sector—reversing the US's downward slide of commercial launches by beating Europe, Russia, and China to contracts. By the time of the Breakthrough Awards, commercial launches comprise 70 percent of SpaceX's launch manifest. (Keep in mind that the bulk of that revenue is coming from NASA—ISS missions reap more money than satellite launches because they include funds for the design, construction, and operation of spacecraft, as well as for the price of a rocket.)

Musk is on a roll, at the top of his game. It's a good time to give him an award. If he can just get to the Hearst Tower in time to save my job.

The interview happened many weeks before, in California. My boss, editor-in-chief Jim Meigs, and I met Musk in Hawthorne, an industrial neighborhood in Los Angeles. Before we arrived, we had breakfast with SpaceX's media coordinator, Katherine Nelson, the same woman who once called me about the whole "violating ITAR" thing. At breakfast I drank two full French presses of delicious coffee as she and Meigs rehashed days when he ran *Entertainment Weekly* and she ran herd for Hollywood stars. This caffeinated indulgence would come back to haunt me.

SpaceX's headquarters is a fascinating place. They bend metal to make rockets, create engines, and have a full staff of young, talented engineers working on amazing new designs. I saw none of that—not even the full-size model of Iron Man, signed by Scarlett Johansson and gifted by Jon Favreau. (It was commonplace in 2012 for people to walk around calling Musk the real life inspiration for the cinematic version of Tony Stark.)

Instead of a factory tour, we made our way straight to Musk's office, a corner desk with no walls on the main floor. On the space beat, you hear a lot about Musk. He has a temper. He pitches wild ideas that teams of engineers must dutifully turn into white papers. He's a megalomaniac.

The guy who sat with us didn't fit any of the above, even though I have since seen how sharp his tongue can be when he's faced with something he feels is stupid. Musk has been a little distant every time I've met him. He always seems a little bored, and when he isn't you can't help but feel that you've accomplished something. Musk is the star football player and the nerd wrapped up into one package, so if you gain his full attention, even for a moment, you feel validated on several levels.

Musk may not devote his full attention to the people around him, but he does have the tendency to see big concepts through to realization. In 2002, he tried to buy Russian surplus rockets to launch a greenhouse to Mars. He found out that it would be easier to build his own—and that, in the coming decade, there would be a market for US-based launches. A decade later, we were sitting in his rocket and spacecraft factory. That shows some kind of steely commitment.

"The probability of human civilization lasting for a long time is much greater if we're on multiple planets," he told us. "There are many things that could destroy life as we know it: natural disasters as well as man-made stuff. But the inspirational reason gets me more fired up. Establishing a self-sustaining base on Mars would be the most exciting adventure I could imagine for humanity. That's the kind of future that I want us to have and I think a lot of people want us to have, particularly Americans."

Like many in the industry, he falls back on altruism and efforts greater than any one person or company. "I think a lot of the American people feel more than a little disappointed that the high-water mark for human exploration was 1969," he said. "The dream of human space travel has almost died for a lot of people. SpaceX is part of restoring that dream."

Musk is good at evangelizing, but he has said this stuff so many times that it seems rote. I asked him if his experience in Silicon Valley have immunized him to risky ventures. After all, he plunked $100 million into SpaceX just to start it up. "I wouldn't say I have a lack of fear. In fact, I'd like my fear emotion to be less because it's very distracting and fries my nervous system," he replied, this time speaking a little less forcibly and off the cuff. "I have this sort of feeling that something terrible could happen, like all of our flights could fail and [Musk's auto company] Tesla could fail and SpaceX could fail, and that feeling of anxiety has not left me, even though this has been a great year. So I feel fear quite strongly; I just proceed nonetheless."

SpaceX offers $160 million launches, about $100 million less than those of Arianespace and ULA. This alone shook the global satellite launch model. "There is no pressure for the Europeans, Chinese, or Russians to bring down their launch costs because they can get whatever they charge," aerospace analyst Marco Caceres once told me. "But now that you've got SpaceX, that's the key to putting downward pressure on the prices. And you should start to see launch costs gradually come down. To me, that's going to be their biggest contribution to the industry."

SpaceX enjoys an advantage over competitors in the way it procures launch-specific supplies. Established launch providers relied on demand from the shuttle program to control costs. But when the shuttle retired, the prices rose. SpaceX is immune to this because it manufactures most of its hardware in-house. I asked Musk about this, and felt a flush of satisfaction as I gained his full attention.

"When I started SpaceX and Tesla, we started off outsourcing almost everything and then over time we insourced more and more," he said. "When you use legacy components you inherit the legacy cost-structure limitations. And you also aren't able to make a product that works together well as a system. If you design the pieces to all fit together in the right way, then it will make for a beautiful result, technically and aesthetically."

I mull this over as I wait for him to arrive to the award ceremony, knowing he will be fashionably late without him realizing that he is actually *entirely* late.

He shows up with his wife, the actress Tallulah Riley, walking up Fifth Avenue as if, well, he isn't a groundbreaking billionaire industrialist. Heading up the escalators with the couple, I see that the crowd has already filed into the auditorium in the Hearst Tower. "Well the ceremony has started," I tell them. "So there's no rush. Should be we get a drink? Maker's Mark is sponsoring this."

"That," Musk says, "sounds great."

In reality, his presence here is a sign of his commitment. He has a launch coming up in less than a week, and every launch (especially for a young company) is vital. His mind, he says, is with the engineers and data sets. Riley proves to be smart and rocket-savvy, which is unbearably alluring. I warn them that there will be an attempt to get them to a post-ceremony dinner, and we engineer a way they can get out of it without them seeming rude.

I manage to smuggle Musk into the theater during one of the ceremony's video segments. Before we go in, he touches my arm.

"So, do I have to make a speech or anything?"

In my head I see an image of my editor strangling me like a Thuggee cultist. "Well, yes," I said. "You'll have the podium when you get the award."

"What do you think they're expecting?" he asks, reasonably.

"I'd say, give them some big vision, Mars landing stuff, something about the permanency of the human species and all that," I say, whiskey-warm and rolling with the moment. "And maybe say something nice about seeing these other award winners. Restoring your faith in the human race."

He nods calmly. The lights go down, and we enter the auditorium. I sneak the billionaire and actress to their seats, smirk at the magazine's scowling deputy editor, and retreat to the bar.

BY 2014, TWO YEARS AFTER THE BREAKTHROUGH AWARDS, THE SpaceX effect is manifesting around the world. French lawmakers responsible for technology investment announce they are backing a plan to speed up the replacement of the Ariane 5. One senator tells reporters that SpaceX's "low-cost launcher constitutes a real and serious threat." The upstart American company is being built from scratch to make money: Visiting SpaceX's facilities, he says, "is like entering Ikea."

People understandably focus on price as the main difference between ULA and SpaceX. But when you see them up close another divide becomes clear. ULA is built for the occasional, important launch. SpaceX is trying to become a steady, regular supplier of space access.

SpaceX president Gwynne Shotwell once told me that her company wants space launch to be "more like an aircraft-type operation. We want to change the flavor of the industry."

By contrast, during a one-on-one tour of ULA's Cape Canaveral facilities, one executive told me that "space launch is a low-rate venture." ULA likes it that way—it's an organization that's comfortable doing ten launches a year.

But playing defense is not an option for the aerospace behemoth. Musk opens a new front in the battle for orbit—launching for the military.

In late 2013, ULA wins a thirty-six-launch contract from the federal government worth $11 billion. Sen. John McCain scrutinizes the "block buy" award and is seeking to review the process. He grills the Air Force Chief of Staff and others over the issue on Capitol Hill. Early the next year, SpaceX sues, saying the process is unfair.

"This is not SpaceX protesting and saying that these launches should be awarded to us," Musk says. "We're just protesting and saying that these launches should be competed." As the only other company able to bid, the statement is true and self-serving in equal measure.

Musk's rockets cost $60 million per launch. With added Air Force requirements, the price rises to $90 million. Still, SpaceX's price tag is better than paying $380 million per ULA launch. Musk wonders out loud, "I don't know why ULA launches cost so much." He cites modern manufacturing techniques and launch procedures as a main driver of efficiency. This is not hyperbole: Designing something from scratch offers definite chances for improvement, whether its friction stir welders in the manufacturing plant in Hawthorne or the rail system at the Cape. Everything that saves time saves money. Anything that trims weight lowers fuel costs.

Despite the cost difference, the aggressive legal move is somewhat of a surprise. They file despite the fact that the Air Force has not cleared them to actually launch, a process that frustrates SpaceX. Musk's company will start competing with the ULA for launch contracts soon, but there are only fourteen opportunities left open to them after the block buy of thirty-six that went to ULA without competition.

They win. SpaceX drops its lawsuit against the US Air Force in exchange for the service making more national security launch missions available for competition. Almost magically, the Air Force quickly determines SpaceX is ready to launch military satellites. Also magically, ULA does not lose a single of the thirty-six launches. Both sides are bound by court order not to discuss the deal, which was struck in a behind-the-scenes mediation presided over by former Attorney General John Ashcroft.

Few will mourn its loss if private space renders ULA extinct. But ULA was born under a cloud, performed better than could be expected, and tried to save money when it could. Now it could become the living ghost of another era.

ULA engineers are talented, dedicated, and experienced. The United States would be grounded without them. But if SpaceX can follow through with its obligations, and can make truly reusable spacecraft by landing and reusing upper stages, then ULA is doomed. It has played the game well, but the rules have changed, and there was no way ULA's hardware and business model can keep up with SpaceX's breakneck innovation.

In 2014, ULA brings in some new blood to help the company adapt: CEO Tory Bruno.

Bruno says his job is "to literally transform the company." Within years he will bring new hardware, partnerships and marketing to a company known for its proudly traditional approach.

He seems like an unlikely foil for Elon Musk. Bruno comes from a background of mechanical engineering, with his name appearing on a few patents that he worked on through the ranks of major military contractors. Prior to joining ULA, he served as the vice president and general manager of Lockheed Martin Strategic and Missile Defense Systems. These firms may be full of stuffed shirts, but they're also staffed with cadres of creative, accomplished engineers.

Bruno has interests outside engineering. He has authored two books applying lessons from the medieval Knights Templar to modern business management. One is called *Templar Organization: The Management of Warrior Monasticism*; the other is *Templar Incorporated*. Applying the lessons that doomed that "multinational organization," he tells the media, can be used to save ULA in its hour of need. "You don't serve the organization, you serve the customer," he tells the *Valley View Star* newspaper in 2014. "As long as you are willing to adapt and do the things that you can do, you get another two hundred years."

His unorthodox nature is fully revealed in speeches and, even more so, on social media. He takes shots at SpaceX, saying they are untested and too risky for national security launches. He also says that the entire cost-reduction attempt—landing empty fuel tanks for reuse—is not ready for prime time. "For the near-term, expendable (rocket flight) is going to be the most practical and cost-effective access to space," he tells the Atlantic Council in Washington, DC.

ULA's long-term reusability plan doesn't rely on empty fuel tanks, but helicopter pilots with nerves of steel. ULA only wants to save the Vulcan engines in the first stage, so their plan is to separate the pair from the rocket after the burn is done. An inflatable shroud, in the shape of a gumdrop, surrounds the engines and they careen back to Earth. A parachute deploys at low altitude, and that's where the helicopters come in. Instead of setting down on the ground using retrorockets, as SpaceX does, ULA's plan saves the weight of the extra fuel by snagging the engines before they even land. The pilot flies past the descending engines, uses a modified hook to catch the parachute lines and ferries the engines back to *terra firma* slung under the chopper. Never ones to rush in, the company says it may be ready in 2024.

Bruno also reaches into the private space world for allies. Once upon a time, it was considered socially responsible to buy space parts from Russia. ULA has long bought rocket engines from Russia, some of the best in the world. But by 2014 the US-Russia relationship had soured, and sourcing those engines start looking like a problem. After all, the rockets they powered carry spy satellites that gaze at, amongst others, the Russians. Diplomats from Moscow have already mentioned cutting off the supply as a way to apply leverage.

Bruno needs a replacement. He finds one in a renegade, space-happy billionaire—Jeff Bezos.

In September 2014, Bezos's company Blue Origin and ULA enter a partnership "to jointly fund development of the new BE-4 rocket engine by Blue Origin. This new collaboration will allow ULA to maintain the heritage, success and reliability of its rocket families—Atlas and Delta—while addressing the long-term need for a new domestic engine."

Bezos gushes in a way that seems calculated to raise Elon Musk's hackles. "ULA has put a satellite into orbit almost every month for the past eight years," Bezos says. "They're the most reliable launch provider in history and their record of success is astonishing."

It's a bold move, and a smart one. Instead of dismissing private space, he is steering ULA towards it to gain some of its innovation, nimbleness, and public luster. Just like embracing the Dream Chaser and partnering with Boeing's Starliner, ULA is proving itself flexible and open to new forms of space travel. The result is seen at the spaceport, where Howard Biegler paints his red Xs and dreams of supporting manned spaceflight.

Bruno uses his social media accounts to help brand ULA and tout its achievements. On Twitter, he establishes himself as someone with comedic timing and a sharp tongue. *AdWeek* notes that he is "surprisingly sassy for an aerospace exec."

After SpaceX nearly lands an empty booster on a barge in early 2015, instead causing it to explode with a crooked landing, Bruno posts a reply with a picture of the Delta Clipper (in shorthand, DCX), an experimental craft that stuck rocket-powered landings in 1993. "Almost. Good luck next time," he writes. "I still have people from DCX. Let me know if we can help ."

The crash is not much of a setback; it's a reminder of how different the approaches to US spaceflight has become. As SpaceX's victories mount in early 2015, with successful satellite launches and cargo missions to the ISS, ULA's road ahead seems to lead uphill. Their best hope seems to be a SpaceX launch malfunction. And before summer ends, they get one.

THE BEST WORD TO HEAR DURING A LAUNCH IS "NOMINAL." THAT means everything is going as planned. Mission control knows this because every modern rocket is streaming data from the sky to the ground, an invisible tether of information. Telemetry and pressure sensors record the flight in real-time. Despite this awareness, when something goes wrong, it goes wrong fast and there's not too much that can be done about it.

The Falcon 9's launch on June 28, 2015, is nominal for 139 seconds. Inside the fairing at its tip are supplies for the astronauts on the International Space Station. There's also a piece of hardware called the International Docking Adapter, which is an extremely complicated airlock. There's a suite of lasers and sensors that communicate with any approaching spacecraft to ensure smooth docking. Any spacecraft parking at ISS needs to have the right connections to swap data and power with the station. The adapter is built to this standard, as will be any other ISS-bound spacecraft, starting with the CST-100 Starliner and Dragon capsules.

At T+140 seconds, the data stream alerts mission control that something's wrong. The second stage, still unlit, is suffering a plunge in liquid oxygen pressure. Observers report a plume of white vapor from the stage shrouding the rocket as it rises, another terrible sign. Within ten seconds the rocket explodes and disintegrates. The Dragon capsule tumbles into the ocean.

The investigation starts before the debris hits the surface, but answers don't come quickly. "Part of reason it's taking a long time is the volume of data, including 3,000 telemetry channels," Musk says.

This data needs to be consolidated and matched up with video feeds to form a single sequence. "When milliseconds matter, it's very difficult to line things up."

SpaceX hires an underwater drone to dive the scene, looking for sunken debris. Musk doesn't hold out much hope it will find anything. "We just wanted to try everything," he says.

The answer is in the hardware. In a pressure-fed engine, a separate gas supply is needed to force fuel and oxidizer into the combustion chamber. In the Falcon 9, helium is the gas of choice to pressurize the propellant tank. High-pressure composite bottles hold the helium that pressurizes the second stage, replacing oxygen that is expended during the launch. They are stored inside the oxygen tanks.

During the launch one of the two struts holding one tank in place broke. The struts are made to survive loads far in excess of what they experience during the launch, so it appears that the part simply failed when subjected to just 3.2 Gs.

Musk says SpaceX is testing every single strut. "This will cause some cost increase in the rocket," he says. "But nothing, we think, will increase the price."

According to SpaceX's data, there are no issues regarding the first stage, which uses the same strut configuration. And the capsule itself could have survived the explosion. "If the software had initiated the parachute deployment, than the Dragon spacecraft would have survived," Musk says. "We're now including contingency software so that Dragon will always attempt to save itself."

SpaceX has lost test rockets before, particularly with the Falcon 1 demonstration flights. But it had never lost one carrying a primary payload. In 2012, the company carried a small Orbcomm communications satellite as a secondary payload during a launch to the ISS, when one of the Falcon 9's engines shut down during launch. That ISS mission was a success, but the rocket released Orbcomm's satellite too low, and within four days it fell back to Earth, burning to a cinder during reentry. Since then, the launch record has been excellent.

Musk says his customers "have looked at all the same data we have

. . . and they agree with our conclusions thus far . . . I'm happy to say none of our customers have demonstrated diminished faith in SpaceX."

Musk says that the string of successes may have had a bad influence on employees who have only seen launch successes. "The company as a whole has become a little complacent over seven years," he said. "This is certainly an important lesson."

The accident shakes up the Cape, as the aura of invulnerability surrounding SpaceX dims. People can say, "spaceflight is hard," but they are not the ones who own satellites or head a government agency that pays to haul its one-of-a-kind equipment into space.

Payloads cost a lot more than the rockets that launch them. Being cheap is not enough—from the customer's point of view, price is not the most important consideration. "High reliability is necessary to be a commercial player," Stéphane Gounari, a space analyst with Northern Sky Research group tells me. "Any new contender will have to prove it to enter the market."

By the time SpaceX returns to flight, it's clear that the company has a fight ahead. ULA is reforming. Arianespace is partnering with Italy and Russia to provide more launch size options. The government is building a massive rocket called the Space Launch System to send people on interstellar missions, bringing a competitor in Musk's quest to set foot on Mars. And the space company owned by Jeff Bezos, a rival billionaire, is emerging from the shadows with ambitions of its own.

CHAPTER 8
TEXAS

On August 24, 2011, a spaceship crashed in West Texas. At first, things went as planned. The pill-shaped craft careened through the air at a steep angle, red-orange jets of flame licking from the bottom. The vehicle shot past 45,000 feet, ripping through the air and breaking the sound barrier.

When things go wrong, they go wrong quickly. The capsule suddenly tipped, instead of heading straight up like a rocket, and the spacecraft becomes more of a missile.

Someone, somewhere made the hard decision to terminate the fight. The engines shut off and, most likely, a line charge of explosives broke open the fuel tanks. The cylindrical vehicle cracked up in a billowing cloud.

Not many people saw it—just a smattering of drivers on Highway 54—but those that did are reminded of the Space Shuttle *Challenger*'s explosion and rain of debris. One driver reported a jet crash. The county emergency dispatcher is dumbfounded.

The closest town to the crash is Van Horn, located about thirty miles away. Ten Van Horn volunteers staff the only fire department within two hundred miles. They responded to the crash, but no one knows what they saw out in the desert.

That's because the spacecraft being tested belonged to Jeff Bezos, the billionaire Amazon executive. It launched from his secret spaceport, outside of Van Horn. "People that go out there are forced to sign a non-disclosure agreement," according to Larry Simpson, publisher of local newspaper the *Van Horn Advocate*.

The locals were pissed. "This thing caught the dispatcher off guard," Culberson County sheriff Oscar Carillo told *Popular Mechanics*. "We were completely left out of the loop. Totally in the dark. And that's a problem."

Days later, Bezos published a blog giving some details about what happened. Even better, he released images and information about the vehicle. Bezos, like others, is working on reducing the cost of spaceflight by building reusable rockets. Part of that effort is flying empty fuel tanks back to a base, landing under rocket power, so that they can be flown again. It turns out that Bezos had been flying various rocket-powered craft from his desert spaceport. The one that crashed had survived several short "hop flights" but couldn't handle a high altitude flight and return to Earth.

The crash lifted the veil of secrecy on Bezos's company, Blue Origin. "Not the outcome any of us wanted, but we're signed up for this to be hard," Bezos wrote. "We're already working on our next development vehicle." And then the veil of secrecy closed once again.

Texas is known for being on the forefront of spaceflight, mostly from the location of NASA's Johnson Space Center in Houston. But an overview of the Lone Star State's space activities reveals a much more complicated picture, of the Texas and of the industry. Instead of large government facilities with long legacies, small towns and small cities are jumping on board, hoping to gain footholds in space.

As a result, mysterious facilities are springing up in remote areas, experimental engines are blazing on test stands, and strange spacecraft are flying through the air. There are more FAA designated spaceports in Texas than any other state.

Just like the early 1960s, Texas is vying with Florida to be the center of spaceflight. This time, however, the struggle isn't for a federal launch center. Now, the state is gambling that commercial entities will operate the spacecraft and paying customers will replace government astronauts. Whether this hope proves to be premonitory or gullible remains to be seen, but the process is already reshaping the Texas economy.

THE COMMERCIAL SPACEFLIGHT INDUSTRY HAS PUT VAN HORN, Texas, on the map for the first time since 1910, when Robert Espy set the world record for roping and tying a goat here. (It took eleven seconds.)

It's still more of a spot of a map rather than a town. The main streets are pocked with empty storefronts still sporting ghostly signs of dead businesses—Rodney's Coffee Café, Fun in the Sun BBQ, Marie's Dress Shop. The main drag's survivors are chiefly hotels, mostly the ramshackle local-owned kind. Away from the oil patches and natural parks, Van Horn is really just a place for tourists to stop if it gets too late to get to the mountains of Big Bend or the caverns of Carlsbad. The state estimates that 10,000 vehicles passed through Van Horn every day. Few ever stop.

To get to Bezos's secret facility, you have to find State Highway 54 in town, turn at the Spanish mission façade of the Capitan hotel, and head north. The terrain is a great example what the newscasters in El Paso call "open country," those nearly empty stretches of desert scrub brush broken by the dramatic, towering Sierra Diablo Mountains on the west and the Delaware Mountains on the east. The highway winds through desert canyons and past a pair of ranches delineated with rusty barbed wire stretched between crooked sticks.

About thirty miles outside of town, one entrance stands out from the rest. The fence is obviously new—metal skewers, no wooden stakes—and the aluminum in the towering light poles still gleams. A video camera keeps vigil, gazing on the card reader–activated gate. A long ribbon of dirt road extends behind the checkpoint, cutting into a basin formed by the surrounding mountains. A handful of buildings, lightning towers, and some sort of squat tank are a thousand yards past the fence, visible but distorted by the shimmer of desert heat.

This is the entrance to a privately owned spaceport. The spaceport has been cobbled together from several ranches that Bezos sought out in the early 2000s. Corporate entities bought the ranches, each with the same address but with differing names taken from famous explorers: "James Cook L.P.," "Jolliet Holdings," "Coronado Ventures," and "Cabot Enterprises." Bezos was behind all of them, and he consolidated the land to create his desert rocket test range.

Blue Origin, not surprisingly, politely declines my request for a tour and interview in 2015. But there are other ways to know what's going on across the fenceline.

Some of the best peeks are courtesy of FAA troves of environmental reviews. Blue Origin's paperwork reveals a network of dirt roads carved in the desert, connecting fifty-foot-high "vehicle processing" buildings with launch pads. The company applied to build a tower for personnel "to access and egress" the rocket at the launch stand, plus a blast wall that could protect workers while they fled, if something went wrong. A guard shack received enough amenities to become a permanent building.

By 2015, Blue Origin has disclosed more details about their hardware. The company is making a launch system called the New Shepherd, named after the Apollo-era astronaut. The New Shepherd is not built to reach orbit, and so the market for flights will be to deliver space tourists and scientific experiments to suborbital space. However, the company is well on its way to design a larger rocket that could deliver satellites and spacecraft to higher orbits.

Blue Origin is capable of launching the capsule higher than 330,000 feet. By the time the company talks about this in public, the capsule and empty booster are ready for soft landings and reuse. His system is the first to launch, land, and reuse rockets that have flown. (SpaceX is the only one to launch and recover in a real mission.)

Bezos has one other trick up his sleeve. He developed the New Shepherd's engine, the BE-4, without any government help. Blue Origin found a willing partner to use the hardware—the United Launch Alliance. The monopolistic joint venture and the market-disrupting startup formally began to working together in 2009 as part of NASA's Commercial Crew Development Program. This gave Bezos a seat at the table of human spaceflight.

ULA had taken a lot of grief over its use of Russian engines, including as a talking point for Elon Musk as he tried to barge SpaceX into the military launch market. The Alliance needed a new, homegrown alternative to power its next generation launch system, the Vulcan. Bezos's engine fit the bill. Development is currently on schedule to achieve qualification for the first Vulcan flight in 2019.

IN DOWNTOWN VAN HORN, THERE IS NO SIGN THAT AN INNOVATIVE spaceflight company has set up shop in a valley outside town. True to the company's quiet approach, the town has no overt Blue Origin signage. Property records show that the company owns several nondescript, one story homes in town.

Blue Origin intern Israel Lopez is standing outside one of them. He drives in to work on projects, but flees back to school as soon as the work is done. "It's really cool to work on things that actually fly," he says.

Mae Mae tends bar at the Hotel El Capitan. An older woman with round face and care-worn eyes, she doesn't shy away from harsh opinions. She's got one about the inadequacies of the new owner of the *Van Horn Advocate*. She will also rail against the influx of drugs in town, and without prompting will sourly note the Sherriff's tendency to shoot off 4th of July fireworks even during a declared "no burn" summer.

But she likes Jeff Bezos and his brother, and the rest of the Blue Origin. "Good people," she says.

It'd be easy to see Bezos as the billionaire who bought Van Horn, the way a feudal lord might claim a barony. After all, Van Horn doesn't really have all that much going for it as a municipality.

Between 1990 and 2004, the region's population decreased, while for the same period, the population in the state of Texas increased. Culberson and Hudspeth counties are both among the twenty-five least densely populated counties in the continental US.

Blue Origin brings jobs to town—good-paying jobs for welders and local construction workers, plus gigs for out-of-town specialists. Everyone is well paid; Mae Mae knows someone who gets paid fourteen dollars an hour to just pick up garbage and sweep up.

Another part of Blue Origin's appeal is that it's entirely funded by Bezos at the local level. In New Mexico, street protestors hold signs railing against the money going into the sputtering Spaceport America project. Here, it's just a guy on a ranch shooting rockets. The company operates thirty miles outside of town, without causing many problems beyond the occasional window-rattling engine exhaust.

Even so, there could be friction between the renegade billionaire and the desert rats in town. But the relationship with Van Horn appears to be important to Blue Origin.

The company has given almost no media interviews, but Bezos himself met with the *Advocate*'s now-departed editor for an exclusive interview in 2005. The people I spoke with in town have a good impression of the way the company conducts itself. When Blue Origin employees act up in town, they get fired. One drunken engineer stripped naked at a local tavern, danced on the bar and police hauled him off, tasered and limp. Blue Origin promptly fired him.

"He knows it reflects badly on Blue Origin," Mae Mae says.[8]

"He" is always Jeff Bezos, and if it seems familiar it's because he's no stranger to the Capitan. The staff there knows Bezos well. This is the place he stays when watching the New Shepard blast off.

Residents know if a test is coming, and how important it is, by the occupancy of the hotel. The big ones are heralded by the arrival of the Bezos brothers.

In October 2016, Blue Origin showed just how transparent they are willing to be. While still keeping media from the test grounds, they maximized social media to build excitement while maintaining strict control over the actual information. The best way to do this is to pub-

8. Discipline is so severe, the rumor has it that even when a spouse acts out of line in public, the employee can be canned.

licize certain dramatic tests. The first was a live stream from Van Horn of a flight of the New Shepherd's emergency exit system, designed to save passengers if something goes wrong during a launch. The capsule has a solid rocket motor that can blaze for two intense seconds, enough time for the capsule to get away and make a parachute landing. In theory, anyway.

"This test will probably destroy the booster," the company said in a statement before the test. "The booster was never designed to survive an in-flight escape. The capsule escape motor will slam the booster with 70,000 pounds of off-axis force delivered by searing hot exhaust."

The test ends better than advertised; the booster survives. Retrorockets slow and steer the craft back to earth, where it settles down on four retractable legs. This is the final test using this launch system, and the Blue Origin staff paint a fifth turtle on the side of the rocket—one for each test flight. The company says they want the hardware to be preserved in a museum.

Bezos is not the only billionaire with experimental, FAA-certified rocket launch facilities in Texas. Elsewhere, Elon Musk is remaking the Lone Star State to be potential epicenter of the space industry.

TO GET TO MCGREGOR, TEXAS, GO TO WACO AND HEAD SOUTHwest for seventeen miles. It's a small town of 5,000 residents, making it small even by Texas standards. At noon on a Thursday, most of the businesses on Main Street are closed for the day.

Luigi's Italian Restaurant, "family owned and operated," is an exception. Inside, most of the tables are filled with lunchtime diners. And at least three of the four-person tables are filed with young men wearing SpaceX badges. Elon Musk's company operates a rocket engine test complex just outside of town.

The SpaceX staff sit in foursomes, many wearing mission shirts emblazoned with the company logo. Their median age is probably twenty-four. Several of them are debating the merits of "creating a mockup" of a rocket engine component, used to convince some higher-up of the design's merits. It sounds like the CAD animation or 3D printing project is in their near future.

"The problem with any mockup," says one, "Is that it has to be all there, done really good with all the details." The other rolls his eyes and flaps his hands, non-verbally telling his peer, "of course."

These engineers are here to work on space hardware; they've earned their mission shirts. Some of the nation's newest engines are put through the paces here, but it's not just the nozzles that get tested. Rocket stages also pass through here, brought by trucks, destined for one of the half-dozen test stands on the facility's property. It also flies experimental spacecraft and is licensed by the FAA as a spaceport. One notice to airmen clears the airspace above the location to just under 100,000 feet.

In town, the space company's stamp is just under the surface of this otherwise quaintly humdrum town. The Chamber of Commerce is closed, but inside a dry-erase board announces a meeting of the town's five biggest businesses, SpaceX among them. The meeting is not public. A brand new Model S from Tesla, Musk's electric car company, zips along the road. Photos of him and local diner owners festoon walls.

Other clues are harder to spot. A colony of twenty-five RVs stand in rows on the lot of the Atria Hotel and RV Park. Welders on long-term assignments stay here. Some of the trailers have laundry hanging on makeshift clotheslines, others have satellite dishes staring skyward. Not all seem occupied; cars sit under tarps and windows remain dark.

"The escort drivers who travel with the rockets always stay here," says Brianna Markum, general manager of the Atria Hotel and RV Park.

The company considers this small town a link in what it calls "an assembly line" that stretches across the nation. SpaceX makes all its engines and thrusters in its Hawthorne, California, headquarters. They're hauled by truck to Texas for live-fire testing, a quality control step that is not optional for any flight-ready hardware. So every space-bound engine comes to this small Texas town. Those that pass make it to Florida's Space Coast.

And frequently—as often as twice a day and sometimes only every other day, Markum says—window-rattling booms shake the town. The sounds are caused by the rush of ignited gasses tearing at high speeds from rocket engines. These rockets generate hundreds of thousands, sometimes as much as a million, pounds of thrust. They make their presence known.

"You'll know when they test," Markum says. "Sometimes the windows will even shake."

Indeed, on Thursday at 11:30 a.m., a deep, whooshing sound fills the air, audible through windows panes and walls. It's a deep rumble, like an airplane engine, caused by exhaust displacing air at high speeds. It lasts at least forty-five seconds before tapering off. A wide plume of white vapor—water turned to steam from exhaust to dampen sound, cryogenic liquid oxygen boiling off as a line is purged, or a mixture of both—rises on the horizon.

The rocket development site, as SpaceX founder Elon Musk calls it, is ten minutes outside of downtown. Getting there is easy, just take Route 84 and turn on Bluebonnet Parkway, past the high school football stadium and schools. The road ends with a manned guardhouse decorated with the company logo.

Beyond the security gate, the horizon is marked with interesting apparatuses. These long, skinny structures are the test stands. There are seven, but you can't see them all from the public side of the road. Engineers mount engines or entire cylindrical rocket stages on these stands to test their performance. One especially large one can host each of the nine engines that carry a Dragon capsule into orbit.

Inside the gates is something that looks like an H.G. Wells death machine from Mars. This is the tallest water tower in America, standing 280 feet high. Its 500,000 gallons of water can be emptied in less than ninety seconds, a cascade that buffers noise and vibration from multiple rocket engines blasting way on a shuddering test stand.

Other tests are even more ambitious. SpaceX is dedicated to landing empty rocket stages to reuse later. They have built rocket-powered craft called Grasshoppers made to jet into the sky, reorient themselves, and guide themselves to make precision landings. These jump from platform to platform on the outskirts of town.

The breakneck pace of testing has provoked a reaction from the local politicians. In May 2016, the McGregor City Council instituted new rules aimed at SpaceX's engine, stage, and low-altitude flight testing—everything that they do. It set new time limits on testing, between 7:00 a.m. and 9:00 p.m.; restricts the length of tests to fifteen seconds or less; and establishes limits on noise.

The punishment is always a fine: "For a test whose volume exceeds 115 decibels, the city requires a permit that includes payment of $5,000 for each test; for a test that exceeds 120 decibels, the city requires a permit and a payment of $7,500 per test; for a test that exceeds 125 decibels, a permit and payment of $50,000 is required," the ordinance says.

This is likely both a response to legitimate local complaints and a naked money grab by local politicians. The word around town is that SpaceX is doing more testing underground, instead of the louder option of mounting the engines on stands. And this shakedown will be tallied as part of the cost of doing business—Musk won't get to Mars doing low power engine tests lasting fifteen seconds.

SpaceX warned the town that things would get a little loud in July 2016. The ensuing test shakes McGregor for more than two minutes.

Astute listeners would know that this burn approximated an actual launch. Keen-eyed locals might have spotted a 156-foot rocket stage wheeled into town earlier in the year, the arrival of the empty tank of a rocket that launched a communication satellite in May and landed on a barge four hundred miles offshore. This test would challenge the stage's nine engines and structure to perform as if new.

These tests are the cornerstone of Elon Musk's dreams of expansion for humanity throughout solar system. His mantra: Reused rockets means cheap access to space, which enables routine, commercial interplanetary flight, and colonization.

These are titan-sized ambitions, and in Texas these dreams either grow big or they go bust. It's happened before, and on the same turf in

McGregor. SpaceX's hard working test area is literally built on the ashes of another launch company, belonging to Dallas banker Andy Beal. Like Musk, Beal launched a crusade to bring the private sector into spaceflight. Only, it didn't turn out so well.

THE RISE AND FALL OF BEAL AEROSPACE IS A CAUTIONARY TALE, ONE that the young engineers at Luigi's probably don't know. Beal's dream was deceptively simple: Form a privately held company to launch satellites into space cheaply without becoming a government contractor.

The disposable Beal Aerospace rockets were designed to be cheap, costing only $25 million[9] but effective, thanks to the combination of their simple old-school design and new lighter construction materials. Its rocket fuel, hydrogen peroxide, was unorthodox but viable.

To do this, one must have a spaceport, a Cape Canaveral for the private sector. Beal endeavored to build his own spaceport somewhere close to the equator, where his rockets could gain momentum from Earth's rapid spin. "If anyone was going to do it, it would be someone like Beal. They took him fairly seriously. He was putting most of his bank profits into the project," aerospace analyst Marco Caceres told me in 2001 when I wrote about Beal for the *Dallas Observer*. "He wasn't wasting a lot of time raising money . . . People thought he was progressing, that this guy could be up and running in a few years."

Beal Aerospace might have done the impossible: Compete against NASA, Boeing, and Lockheed Martin—and win. He wanted to convert Sombrero Island—a ninety-five-acre rock about a hundred miles from St. Croix, belonging to the island nation of Anguilla—into a spaceport. There are no trees or shrubs on Sombrero, and the Dutch had long ago strip-mined Sombrero Island for it trove of bird guano. The mining and severe weather erosion flattened the only parts that peaked above sea level, so the "hat" it's named after now just looks like the brim.

The company stumbled into unwanted controversy from its de-

9. NASA pays $50 million to $80 million for a comparable launch.

cision to build a headquarters and manufacturing plant in St. Croix in the US Virgin Islands. The rockets would be assembled and fitted with customer satellites before being shipped via barge to Sombrero for launch. Environmentalists and historic preservationists didn't appreciate its location at a slave plantation, and protested. A judge ultimately ruled that the land had been donated for use as a public recreational area, thus precluding any large-scale commercial activity.

Other obstacles arose. The Venezuelans claimed ownership over some of swampland Beal had eyed for his assembly facilities. Their diplomatic tantrums cause delays, headaches, and a dim view of the project from Washington, DC. Besides, Boeing and Lockheed Martin were going at each other like gladiators in the 1990s, fighting for the hundreds of billions at stake in national security launches. No one in NASA or the Pentagon wanted to back a dark horse candidate.

On May 20, 1999, and Beal pleaded his case before Congress. "Please, please do not give companies billions of our dollars to play around with experimental programs," he told the Senate Committee on Commerce, Science, and Transportation. "You will create jobs by spending public money, but you absolutely will not produce low-cost commercial access to space."

His words were prophetic. Boeing and Lockheed Martin's competition ended in the fifty-fifty partnership called the United Launch Alliance, whose products are decidedly not low-cost. Andrew Beal's ambitious idea to use new tech to carve a piece of the commercial launch market echoed the twenty-first-century schemes of Blue Origin and SpaceX. But he missed one crucial aspect that is key to cheap spaceflight of the current entrepreneurs: reusable hardware.

Beal gambled and lost, but others have built on his failures. The defunct aerospace company left test stands on the outskirts of McGregor, just off the Bluebonnet Parkway. Elon Musk, ever with an eye for savings, saw it as a prime spot for his new commercial space venture. In 2003, SpaceX set up shop and invested $50 million in renovations over ten years.

Musk moved into Beal's rocket test center, and inherited his dream. But instead of an island off the coast of South America, Musk has Brownsville, Texas.

GIL SALINAS WAS HOSTING HIS TEN-YEAR-OLD DAUGHTER'S BIRTHDAY party in 2010 when the phone rang. He picked up the receiver and heard a pitch from state officials. It would literally change the life of the young executive vice president of the Economical Business Development Council in Brownsville, Texas. Elon Musk was considering building a spaceport in Brownsville, on an isolated stretch of Boca Chica State Park bordering the Gulf of Mexico.

Salinas is a well spoken, plain dressed guy, cue-ball bald but sporting some diabolically sharp eyebrows. With that phone call he and the city he represented were propelled into an epic attempt at interplanetary exploration.

Within two weeks of that first call, Salinas and a delegation of South Texas officials met with Musk himself at SpaceX's headquarters in Hawthorne, California. "That's when he told us about his vision," Salinas says.

Musk is very good at articulating this vision. Anyone who's sat across from him and heard it face-to-face agrees it's compelling. Musk doesn't couch the argument for his company's mission in returns on investment. He speaks about using an innovative company to make spaceflight cheaper, opening access to space. The end goal is not market dominance; the goal for Musk is human spaceflight to Mars. When he gets humanity off one fragile planet, he'll ensure the survival of the species. Every step, every contract, every test launch, every government mission at SpaceX is part of this goal.

Musk asked to use Brownsville as the literal springboard to Mars. He wants to fund a colony on another planet. Salinas left the meeting and found himself having to evangelize Musk's vision to his fellow South Texans.

"They thought we were nuts," he says.

Salinas agonized as the negotiations unspooled. The city produced a series of environmental reviews and incentive proposals. They lobbied the state to loosen up laws on beach construction to entice the company. Through 2012 SpaceX very publicly considered Georgia, Florida, and Puerto Rico for locations of a built-from-scratch spaceport, which ensured a healthy competition of subsidy packages between contenders.

In 2013, officials in Brownsville created the Cameron County Spaceport Development Corporation. On the other end, Texas state officials created the Spaceport Trust Fund and promptly routed $13 million to the county's development corporation. These moves landed the deal, much to the chagrin of other would-be locations.

Salinas says he didn't know what SpaceX would do until the few moments before the announcement. At the time, Musk said his company planned to invest $100 million into the launch site in Boca Chica State Park within three or four years.

The FAA gave SpaceX authority to build in July 2014. As it turned out, that was the easy part. I visit in 2015 for a firsthand progress report.

The directions to get to the spaceport are pretty simple: Pick up State Highway 4 in Brownsville and drive east until you hit sand. The battered two-lane highway takes gentle curves, skirting protected nature preserves and the US-Mexican border. There are signs marking this remote scrubland as the location of the last battle of the Civil War (fought long after Appomattox) and a Border Patrol checkpoint.

Near the end of the road stands an empty water tank with a handwritten greeting, "Boca Chica Village Welcomes You." But the handful of inhabited homes here that are not boarded up have tall fences, CCTV surveillance signs, and German Shepherd guard dogs. There are large water tanks (like the one at the entrance) at each; Boca Chica Village has no utility services.

Highway 4 ends with a sign that warns: "Pavement ends 1000 Feet." The weathered asphalt doesn't stop as much as it seems to be erased by creeping sand. There is an RV and a Jeep parked on the shoulder. A pale, bearded face peers from the RV window, but no one emerges.

It's hard to see Musk's dreams from this barren, wind-whipped patch on the Gulf of Mexico. Well-sculpted dunes rise on both sides, overgrown with beach grass. The beach itself contains nothing but rusted metal garbage cans and signs warning people not to run over sea turtles.

The Border Patrol agents at the checkpoint say that no work had been done since the official groundbreaking ceremony. They note that the sign erected for the groundbreaking ceremony in September is no longer standing. "They put it up before the press got there," one agent says. "As soon as the event was over, they took it down."

The sort of early work done in 2015, which includes surveying, design work, and environmental reviews, leaves little trace on the beach. Called for a reaction to local talk of delays, SpaceX maintains that there was never any schedule, so there can be no delay.

Despite a lack of tangible progress and some restlessness in the region, Brownsville has already reaped the benefits of the nascent spaceport. After the space company inked a deal, other companies started calling. "Being able to say we work with SpaceX opens doors," Salinas says. "They say, if an innovative investor like Elon Musk is coming here, there must be a reason. The economic development pipeline is now open."

There are more direct economic boosts as well. A Michigan company and SpaceX supplier, Paragon, bought a local tooling company so they could supply parts to the new launch center. Salinas says museum operators and hotel chains have been calling, anticipating tens of thousands of visitors to watch each launch. And the airport is rebuilding its terminal and doubling its runway length to accommodate massive cargo airplanes carrying satellites.

Other local institutions stand to gain from the SpaceX effect as the spaceport takes root. University of Texas-Brownsville's astrophysics department was not even ranked by those who measure such programs. But then came Elon. As the SpaceX deal really started shaping up, the Greater Brownsville Incentive Corp. forked over $500,000 to the university to start a spacecraft tracking program called "Spacecraft Tracking and Astronomical Research into Gigahertz Astrophysical Transient Emission." (Yes, STARGATE.)

At the spaceport groundbreaking, Governor Rick Perry announced STARGATE was getting a windfall. The state committed $4.4 million, and the university ponied up $4.6 million, to the program. But these investments existence depend on SpaceX, and the Memorandum of Understanding between UT and SpaceX to support the program. The most tangible symbol of this dependency: STARGATE will be housed in SpaceX's command center on the beach.

For a while anyway, SpaceX spokesman Phil Larson insists the spaceport construction is on track for 2016 launches. After my 2015 visit and an article on the lack of progress, he makes it a point to complain to me for the perceived skepticism. It takes nearly a year for the company to admit that the spaceport would not be open before 2018. By then, Larson is no longer working there. SpaceX seems to chew up and spit out media relations types.

In early 2016, trucks start to arrive, hauling enormous amounts of dirt. SpaceX tells the *Brownsville Herald* that they plan on dumping 310,000 cubic yards of soil onto the property to create a stable foundation on which to build a launch pad. That's a hundred-yard brick of dirt standing thirteen stories tall, the newspaper calculates. The increase in traffic batters State Highway 4. Patch crews from the state DOT are deployed regularly to fill the new potholes. By the start of 2017, SpaceX is still piling dirt on the beach.

Texas's courtship of spaceflight is not limited to SpaceX. The state leadership wants the space industry to come to here as much as the local communities are eager to host them. The result is a textbook example of how public money is diverted to seed the private space industry.

IN MID-DECEMBER 2016, SPACEX'S DIRECTOR OF GOVERNMENT relations rattles the cup at the state of Texas. A joint legislative committee hearing in Brownsville, Caryn Schenewerk complains that the state of Texas didn't appropriate any money to its Texas Spaceport Trust

Fund. Her company, owned by Elon Musk, is the largest beneficiary of the fund.

"Unfortunately, the Spaceport Trust Fund was not funded in the 84th Legislature," she testifies. "We will certainly be advocating for it to be considered by the 85th." She didn't hesitate to call out Florida's support of spaceflight, which has an infrastructure fund that hands out grants worth nearly $20 million each year.

In 2013, the 83rd Legislature appropriated $15 million to the Spaceport Trust Fund, but by 2016 the fund is dry. "All of the appropriated funds have been awarded and no funds are available at this time," says Sam Taylor, deputy press secretary at the Governor's office.

The Spaceport Trust Fund is a good example of economic development or corporate welfare, depending on your point of view of these things. It's a process, not a competition. The first step is to create a development corporation that can solicit and accept taxpayer money.

SpaceX gets the bulk of the cash, but the state sends the remaining $2 million to the Midland Spaceport Development Corporation. Midland is a center of West Texas commerce, an oil town that anchors the region. The FAA granted Midland International Air & Space Port its Commercial Space Launch Site license on September 17, 2011, making it the first primary commercial service airport to be given a spaceport designation.

The airport places a symbol of its brave new future—a model of its marquee spacecraft tenant, XCOR's Lynx—in the airport's lobby. The Lynx looks like a winged, fully reusable, suborbital space vehicle that is designed to safely carry two persons or scientific experiments to the edge of space and back up to four times per day.

XCOR is the anchor tenant of the spaceport. Midland Development Corp. has allocated $11.5 million for XCOR's move to Midland from Mojave. The deal included a $1.5 million refurbishment of the XCOR hangar.

The Lynx is a pretty cool idea. Starting from a runway, it is designed to corkscrew its way to suborbital space, and then land like a plane on the same runway. It would be aimed at tourists, and the ride would cost $95,000 per ticket.

But the Lynx won't be flying from Midland. That display model is as close as the airport will see to the spaceplane. In May 2016, the company announces a change in executives, layoffs, and the indefinite delay on all work on the Lynx. Instead, the company plans on focusing on its rocket engines, which are part of a competition to supply ULA. Engine testing is moving from Mojave, California, to Texas, but the FAA requested that engine testing be moved away from the airport.

"It makes total business sense," said J. Ross Lacy, Midland city councilman and a member of the Texas Aerospace and Aviation Advisory Committee, as well as the Midland Spaceport Development Corporation Board. "They made us aware of the layoffs and called us to update us on their plans."

Part of the reason why Lacy is not howling mad is that Midland already got a bump from landing XCOR as a tenant. Along with the flight company came Orbital Outfitters, which builds flight suits. Orbital Outfitters will get to use the Midland Altitude Chamber Complex, a structure built by the spaceport with $3.5 million of taxpayer money containing equipment hard to find outside of NASA. It's set to open for business in 2017, Lacy says, with seats for ten and two people, as well as a chamber for instruments.

Other space companies are also eyeing Midland, he says, and deals are pending. "It took XCOR to get a foot in this industry," Lacy says. "Things are still moving along. We're glad they're here."

Things may have been moving along when we spoke, but now XCOR looks shakier than ever. In 2017, the company lost the contract to work on engines with ULA, a revenue hit that prompted layoffs in California and Texas. All the employees in Midland have been axed, except for three contract workers; eleven people in Mojave also lost their jobs.

Company officials in July scrambled to reassure Midland public officials that the Lynx spacecraft is still viable—and tell hundreds of people who bought early tickets the same thing. But the spacecraft now looks more like an asterisk in a space history book than anything else.

Even as these successes and failures continue to play out from the previous round of financing, Texas politicians are being asked to refill the Spaceport Trust Fund. SpaceX's plea for more money has been echoed by the Aerospace and Aviation Advisory Committee, a panel of advisors selected by the Texas governor. They've officially asked Texas to once again restock the coffers to promote Lone Star State spaceports.

This recommendation is not surprising, given the makeup of the advisory board. Taylor says the Aerospace and Aviation Advisory Committee "includes representatives from the aerospace industry."

Considering that many spaceports have been prompted into creation by aerospace companies, it's a fair bet that these economic development folks will solely advocate for public funds to be produced. And they are not apt to get much pushback from the other board members, who also stand to gain by steering public funds to infrastructure that supports their companies' bottom lines.

The advisory board is rounded out with executives from Lockheed Martin, Bell Helicopter, Gulfstream, Raytheon, and Boeing. These players have definite interests in promoting aerospace capabilities at airports, since many of these companies already have operations.

Indeed, the advisors include many who will receive the money they are advising be spent. For example, advisory committee member J. Ross Lacy is the president of the Midland Spaceport Development Board, the recipient of $2 million from the state fund. Gil Salinas is the executive vice president of the Brownsville Economic Development Council and a key supporter of the SpaceX spaceport. Bob Mitchell is the president of the Bay Area Houston Economic Partnership, which has made the Ellington Field spaceport one of it's official "economic development issues."

It's a delicate balance between the past and the future, between preparation and bad investment. Texas's spaceports may not each thrive, but collectively they are a symbol of the state's faith in this aerospace movement.

"Building a spaceport is not an easy thing, believe me," says Arturo Machuca, general manager of Houston's Ellington Field Spaceport, at a 2016 conference. "We will get there as soon as industry gets there. Whenever they are ready to fly we're ready to host them."

CHAPTER 9

WACO AND TUCSON: NEW ARRIVALS

THE JOHNNY APPLESEED OF SPACEPORTS HAS A NEW CLIENT. THE Texas State Technical College (TSTC) and Greater Waco Chamber of Commerce want to convert the school's airport into a spaceport. They need spaceport consultant Brian Gulliver to get it done.

The Waco Chamber of Commerce and TSTC hires the Kimley-Horne consultant in 2016 for nearly $200,000 to analyze what infrastructure is needed to accommodate spaceflight.

Gulliver takes the study in pieces, and the first involves fieldwork to determine if the airport is physically ready for spaceflight. In late December, a draft of the study finds that Waco's infrastructure could handle the operation of airplanes that launch space rockets, air-launched spaceplanes, or spaceplanes that can use their own onboard engines to blast into space.

It's the start of a three-year FAA application process. Now comes the hard part: convincing the local airspace managers, state officials, and the other tenants at the airport that having a spaceport in Waco is a good idea.

The twenty-first-century spaceport movement is molting, yet again. First, spaceport dreamers envisioned standalone facilities, like

Spaceport America. Then, municipal airports teamed with prospective tenants to be cleared for spaceplane launches. Now, aerospace dreamers are proposing facilities that break every mold. Waco is turning the idea of a tenant-driven spaceport on its head. Similarly, in Tucson, industry veterans are putting the entire idea of rocket-based spaceflight to the side in favor of balloons.

The plans are different, but the ethos is the same. The concept of what makes a spaceport is still changing, and still bringing more communities and people closer to spaceflight than ever before.

BRIAN GULLIVER WAS BORN WITH SPACEFLIGHT IN HIS BACKYARD. Growing up in Orange City, Florida, between Orlando and Daytona Beach, he had an active spaceport as the backdrop to his childhood. He remembers his classes at Deland High School marching to watch Space Shuttle launches and hearing them during Little League games.

The allure of mechanical engineering led him to the University of Florida and University of Central Florida, and then on to the engineering firm of Reynolds, Smith & Hills working on contracts for NASA. Decades after gaping at the shuttle as a kid, he finds himself working on ground infrastructure at Cape Canaveral.

Gulliver designed lightning rods on SLC 39B at Kennedy Space Center. He also redesigned a room within the access arm of the tower and Space Shuttle, "the last room the astronauts are in before they board."

His career took him to spaceports from coast to coast. "I've worked at both Wallops Flight Center, Pad 0A concept development and design, and at Vandenberg Air Force Base's SLC-3E design," he says.

The rising number of spaceports and a private space boom buoyed Gulliver's career. Until aerospace firms started creating new spaceplanes, everything Gulliver designed for new spaceports supported rockets rising from launch pads. But during this first phase of the private space movement, the commercial space world remained fixated on the familiar style of launches. "The idea was a bunch of smaller, vertical lift pads," Gulliver says. "Like mini Kennedy Space Centers."

But that soon changed as more companies joined Virgin Galactic in commercializing spaceplanes and in building a new generation of air-launched rockets. In 2010, Florida made a move that opened the eyes of many smaller airports with big ambitions. Officials with Space Florida, the state's official economic development agency, knew that spaceplane programs at the Kennedy Space Center would need a home after they left the cradle of NASA development. The state agency wanted to make sure that there were runways available in case these projects ever get off the ground.

They contracted Gulliver to navigate the process. With him, Cecil Field (near Jacksonville, Florida) received its FAA designation as a spaceport. There, the airplanes would fly over the open ocean, drop the space rocket, and veer away as its engines ignited.

Gulliver saw that this process could be repeated. "After the Jacksonville project was done, we saw there was a market emerging to help spaceports co-locate at existing airports," he says.

He hit the road, pitching the idea to airports by presenting papers at space conferences. "While commercial spaceports that support vertical launches are still being developed, the largest growth has been in spaceports designed to support the horizontal launch sector," reads one, presented in 2012 at the AIAA Space Conference in Pasadena. "In addition these new commercial spaceports are being developed in more diverse locations around the country and will soon be more numerous than traditional spaceports."

Gulliver left his job to become the "Spaceport Practice Leader" for the firm Kimley-Horn. Over the years, he has found steady work creating spaceports across the United States and abroad. He has become a mercenary apostle of spaceports, quietly planting the future of space travel from town to town.

IT'S THE FIRST WEEK OF JANUARY, THE WEEK BEFORE CLASSES BEGIN AT Texas State Technical College in Waco. The campus is post-holiday sluggish, but shows signs of renewed life. The bookstore is open, but the staff outnumbers the shoppers. Signs warning of late registration fees are posted around the student center, but almost none of the nearly 6,000 undergrads who attend are around yet to read them.

TSTC focuses on two-year training in practical, professional trades. The dozens of disciplines taught here read like a register of steady-demand skilled labor: electrical line workers, refrigerator techs, dental assistants, land surveyors, landscapers, and computer system engineers.

The disciplines are pretty bread and butter, but TSTC has something no other college can match: the largest airport owned by an educational institution in the United States. TSTC airport dominates the eastern fringe of the campus, complete with two runways and an air traffic control tower. It's a general use airport, and anybody flying by can radio the tower and request to land there. A subset of students here aims at niche aviation work, such as avionics repair, aircraft hangar maintenance, and air traffic control. There are also industrial tenants at TSTC and mammoth hangars, some with doors towering more than sixty-five feet high, indicating the size of the airplanes that come here for maintenance work.

Today, three white jet planes are wheeling over the airfield in wide oval, racetrack patterns. They land and takeoff again without fully stopping, what pilots call "touch-and-gos." Inside the cockpit are Air Force trainees getting practice time in twin engine T-1 Jayhawks. These are the planes that airmen fly to learn how to operate larger cargo haulers and aerial refueling tankers, and TSTC is a favored place to practice approaches and stop for lunch.

The airfield sees an impressive 101,000 takeoffs and landings a year, and hosts a major airshow every year. Air Force One landed here when George W. Bush was president, ferrying him to his Crawford ranch. Not bad for an airport that doesn't even have its own radar.

Kevin Dorton, the airport's manager, watches the T-1s from the catwalk atop the ATC tower. Then he points in the other direction, to a hanger and several gray, 167-foot-long airplanes. One of the airplane's propellers is spinning. "I think those are P-3s with the Royal Australian Air Force," he says, not entirely certain of the identification.

He's more certain when it comes to the funding history of the hangars the P-3s are using. In 2014, the airport assessed that the World War II–era building needed $4.4 million in modernization in order for P-3 military aircraft to be maintained here. When airports stare at such bills, they turn to the myriad places to help defray the cost. Targeting and collecting public state and federal cash is a foundational part of running an airport.

Dorton would have gone to the Texas Department of Transportation for grant money, but those funds are reserved for routine repairs and improvements. The school turned to local government, and the city and county provided $1.1 million under the banner of economic development. The company that rents the facility, L-3 Technologies, chipped in $900,000. TSTC got a $1.8 million low-interest loan from the governor's office and paid $600,000 out of pocket.

Airports are often staffed with aviation fanatics who have one eye fixed on the airspace above and another begrudgingly focused on spreadsheets on the ground. Dorton, a former bank auditor and moonlighting rancher, is not one of these. Ask him if he wants to learn to fly and he plants his legs a little wider and says, "I want to stay firmly on the ground."

But he does have passion for the facility, though, and is no mere pencil-pusher. When no one else can, he checks the runway for debris and at least once cleaned a pile of coyote crap from the runway with his bare hands to expedite a flight. But his approach to problem solving is admittedly pragmatic.

"I hate to sound like a reformed banker," he says. "But I think of things as black and white, in terms of return on investment. Here, ROI is anything that helps us with the mission. So a tenant that will make infrastructure improvements is one return. Having a place where our students can get internships or jobs, that's another kind of return. It's not just cash flow."

Dorton may seem like an unlikely proponent of a high-risk, almost science-fiction direction for this trade school airport. But he is one of the prime movers trying to enable TSTC to become a spaceport, a place able to launch and land spacecraft from its runways.

Next week, as the semester's classes start, Dorton will head to Fort Worth to meet with state and Federal Aviation Administration officials. They will look over the plans and advise the group on any airspace conflicts that could stop the show. "Sixty percent of the feasibility study is done," Dorton says. "The other 40 percent is pretty much negotiating and navigating the landmines in the sky . . . We have to work with American Airlines and Southwest flights."

If the FAA doesn't throw any monkey wrenches into the plan, the feasibility study will be done and the spaceport effort would enter the next phase, the environmental study. If all goes well, in less than two years TSTC's spaceport could be cleared for launches.

DORTON, EVER TRUE TO HIS BANKING BACKGROUND, PROCESSED the idea of a spaceport in Waco through his usual risk/reward filter. That meant getting a reaction from the biggest tenant at TSTC airport: L-3 Technologies, a defense contractor based in New York City.

L-3 is Waco's largest industrial employer and is what Dorton calls the airport's "900-pound gorilla." More than 1,000 employees work at the company's Platform Integration facility. "Platform" in this context usually means airplanes and "Integration" means installing new equipment onto existing aircraft, like 747 airliners, and a slew of large military cargo planes, like the C-130 and C-27J. They also work on Navy sub-hunting airplanes like the P-3.

The work L-3 Platform Integration does in Waco has been waning as military contracts withered. This year the company laid off 120 employees in Waco. These losses come on top of 314 reductions made since 2014. This can be attributed to the normal cycle of defense contracts, but it also makes the company eager for new work. There is one business sector that L-3's Waco facility has experience with that is expected to grow: spaceflight.

In March 2016 *Cosmic Girl* landed at TSTC and parked inside one of L-3's big hangars. Richard Branson's company, Virgin Galactic, bought the 747-400 to convert the airliner into a flying launch pad for small satellites. The company, founded with a focus on space tourism, is now also chasing future customers who want to launch satellites.

Virgin chose Waco for more than its experience with 747s; the Platform Integration staff has worked on previous space programs. The SOFIA airborne observatory—a 747 heavily modified to support a twenty-ton far-infrared telescope—is the best-known success that has roots in Waco. "These programs are not frequent, but complement the other work we do," says L-3 spokesman Lance Martin.

The *Cosmic Girl* project will be done well before the spaceport designation comes to TSTC's airport, but the work could be a harbinger of future markets for L-3 within the still-nascent private space industry. The contract would be bigger if the converted 747s could conduct test launches from Waco, for example.

"For L-3 Platform Integration, any kind of space-related work that involves horizontal launch would have potential," Martin says. "Our aerospace engineering and flight sciences expertise would be particularly important for these types of programs, so a spaceport designation would add an avenue for future growth for our division . . . We did not approach the Chamber about this idea, but certainly support it as it can benefit the entire community."

There are other aviation companies that could benefit from the spaceport designation, but none of them are as large as L-3. Absent other tenants with spaceflight aspirations, the hard-won designation could become an unused, expensive piece of paper.

No one at TSTC sees SpaceX as a tenant. The company operates an engine test facility in the county, but SpaceX's proximity has no bearing on the spaceport. Company officials tell the *Dallas Observer* that they don't have plans to use the spaceport at all, since they launch vertical rockets, not spaceplanes, although SpaceX supports the idea (as a member in good standing with the Greater Waco Aviation Alliance) and company officials have toured the TSTC.

Having a spaceport in the backyard could benefit another Waco mainstay: Baylor University. The university operates its own industrial park called the Baylor Research and Innovation Collaborative (BRIC). L-3 rents lab space there, working on advanced composite materials and niche aircraft communications research.

The university, county, and aerospace company see these research parks as win-win-wins. In 2014, L-3's president of L-3 Platform Integration, Nick Farah, spoke to tech school students about the BRIC. "I want to attract innovative engineers," he said. "I want to be at the heart of technological advancements. You help me attract talent. At the same time, together we are helping the local economy by attracting and retaining graduates."

How this looks with a spaceport in the mix is hard to predict. In an ideal situation, L-3 would win a contract to develop a space-related project, installing components and testing them during actual launches. This work would spill over into the BRIC, where engineering students would be able to get their hands on flight-ready space hardware. When that space-related contract ends, L-3 could have an experienced team, including newly hired students, in place to bid on another contract.

These benefits may not manifest right away, or ever. But the spaceport designation, that hallowed piece of paper from the FAA, makes this type of job possible.

Terry Stevens says the very existence of the TSTC is a testament to ambitious, far-seeing planning. "When they shut down Connolly Air Base in the '60s, it could have been a disaster," he says. "But turning it over to become a TSTC school, that was brilliant . . . It's hard to plan ten or twenty years out, but that's what leadership is."

Stevens is a member of the State of Texas's aerospace advisory board, which is advocating that the legislature refill the Spaceport Trust Fund in 2017's session (see the previous chapter). Dorton also makes a point to say that founding a spaceport "opens up other sources of revenue from the state."

One of the first steps to get that revenue is to create a development corporation. To put this in spaceport terms, a development corporation is a vehicle that a municipal entity must create to collect state grants from Austin. In March 2015, county commissioners approved the creation of McLennan County Spaceport Development Corporation.

If the Texas State legislature refills the Spaceport Trust Fund, this tech school wants to make sure they are ready to get the money.

THERE IS A SHARED DREAM THAT YOU ARE BOUND TO HEAR ABOUT while hanging around airports-turned-spaceports. All of them want a piece of the coming "point-to-point" space travel market—airplanes that take off, reach suborbital altitudes, and then land across the globe in a matter of hours.

Let's say an investor in Dubai wants to come to New Mexico for a meeting with Ted Turner. This well-heeled fellow would board a spaceplane, a "single stage to orbit" vehicle that doesn't need a mothership or rocket to whip into space. The weightlessness is just a perk, or a curse, depending on the traveler's point of view. The investor certainly would gawk at the curve of the earth from a window seat. The landing at Spaceport America is a long, fast glide and a touchdown on the long runway.

The spaceplane itself can be refueled and is ready for a return trip, perhaps carrying microgravity experiments, cargo (let's say, chilies from the town of Hatch), or other passengers. Most likely, the spaceplane waits for the customer to finish business and returns to Dubai.

Any spaceport that can accommodate spaceplanes could become a destination for these high-end, globetrotting flights that the rich will covet—cost be damned. The military also likes the idea for quick logistics, as do commercial shipping firms like UPS for the potential efficiencies.

But there's a more benevolent view, one where spaceplanes spread the opportunities of spaceflight to more companies, universities, and the public. I get the best version of this future from Mark Sirangelo, corporate vice president and head of SNC's Space Systems. That's the company making the Dream Chaser, the spaceplane that takes off on the tip of a rocket but lands like an airplane.

If point-to-point space travel ever becomes a reality, the network that Gulliver is founding will be the nodal points of such routine space travel. If this future doesn't come to pass, these unused spaceports will hardly be the first totems built for transportation schemes that never came to fruition. There are mooring stations for zeppelins on the Empire State Building, after all.

On January 9, 2016 classes start again at TSTC. Students spring from on-campus buildings, the walkways become jammed with foot traffic and the cafeteria, staffed with campus-based culinary arts graduates, reopens. This is good news for the guys in the air traffic control tower, who missed the close proximity to what one of them calls "the best burgers in Waco." It's also positive for visiting pilots, who can visit the cafeteria during training flights instead of driving for fast food in the airport's courtesy van.

That Monday afternoon, ninety miles north, Kevin Dorton and Brian Gulliver enter a conference room in the FAA's Southwest Regional Office in Fort Worth. There are eight FAA officials in the room, and another ten or so on the conference call, many of them managers of air traffic control systems across Texas. One NASA official in Houston sits in on the call.

Dorton's plan is to let Gulliver, "our hired gun," take the lead.

Gulliver opens the meeting with the obligatory PowerPoint presentation. The slides show various spaceplane designs, flight paths of planes and rockets, maps of where the spaceports are, and primers on FAA spaceport licensing. But the meat of the afternoon's gathering is to discuss how a space launch operation could coexist in central Texas's airspace, along the Interstate 35 flight corridor connecting San Antonio, Austin, and Dallas-Fort Worth.

Every meeting is critical in this process, even an early one like this. For Dorton, it's the first time he will hear of restrictions and limitations to when and how his spaceport can fly. For Gulliver, it's a chance to demonstrate how his earlier experience benefits his client. His firm will bid for the contract to do TSTC's environmental study, the next big step after the feasibility study is done.

The FAA's concerns are clear. Any launch would have to coexist with regional flights between cities and longer-haul airliners passing through. The real issue is weather: Airlines reroute aircraft to avoid weather systems, often pushing them over Waco. With every minute in the air, those planes burn fuel and take money away from an airline's bottom line. Anything that takes landing or routing options away would be met with resistance.

But this potential airspace conflict is not a deal killer, Dorton says later. "All this was expected," he says. "We may have to restrict our launch times between 7:00 a.m. and 10:00 a.m., or launch straight east or west." Furthermore, air-launched rockets are more flexible when it comes to when and where they operate. "You can just take off and get into a pattern, meshed with ATC, and then go vertical when it's clear . . . We're OK with that."

The meeting sounds like a win for Gulliver, too. Like a good hired gun, he doesn't comment on work he does for clients, but Dorton said he was pleased. "He clearly had a rapport with them," he says. "They knew who he was and his earlier work. For our part, we were well served."

The meeting lasts an hour and a half. The path ahead looks like meetings, studies, negotiations, conference calls, check-ins with state politicians, loan applications, zoning plans of a proposed ten-acre industrial park, and checks to consultants. The back end of the barnstorming and bravery of experimental flight is bureaucracy and business. For Dorton and the rest of the Waco spaceport dreamers, it's all the same. It all just looks like the future.

TABER MACCALLUM RECLINES IN THE OFFICE CHAIR AND MOTIONS at a video screen. On it, a team of engineers is lashing a Google executive facedown under a high-altitude balloon. The video is from 2014, two years ago, when MacCallum was part of the group involved in a scheme to drop multimillionaire Alan Eustace from the edge of space.

The video says as much about MacCallum as it does about World View Enterprises and its future here in Tucson, Arizona. Now the gang that accomplished this feat wants to send balloons to the edge of space from a newly minted, one-of-a-kind spaceport on the edge of town.

MacCallum is one of the original eight crewmembers of Biosphere 2 in Arizona. He spent two years in that that sealed environment. Later, he founded a company called Paragon Space Development Corporation that designs life support equipment for extreme divers and astronauts.

In 2011, Eustace sat down with MacCallum to discuss ways to loft him to the stratosphere under helium balloons and then free-fall back to Earth. Anyone else would have been shown the door, but the multimillionaire had the money and, as Google's "Senior Vice President of Knowledge," he had geek-community, rule-breaking pull.

Even someone greedy would probably laugh and walk out of the room. But in 2011, the seed of confidence that marks private space's rogue aeronauts had already started to grow. MacCallum and his partners at Paragon—fellow Biosphere alumni and early commercial space industry boosters—proved up for the challenge. Eustace was offering a project that exemplified the attitude that private firms can be capable of staggering aeronautical feats. So MacCallum said yes.

Working in secret, the team procured the balloon, parachutes, oxygen chamber, and pressure suit required to make the twenty-six-mile-high drop possible. Paragon created the life-support system that enabled him to breathe pure oxygen during the trip.

They launched him from an abandoned airfield in Roswell, New Mexico. The trip up to 123,334 feet lasted two hours; it took fifteen minutes to come down. The free fall broke the sound barrier. Eustace reached more than 820 miles per hour before opening his chute and drifting back to the planet's surface, seventy miles from where he started. Eustace still holds the record for highest free fall jump.

"It wasn't seeing the curvature of the Earth that made the biggest impression," MacCallum says. "He says that the slow ride up gave him this perspective."

At the time, someone from the US Parachute Association community, on hand to certify the flight, made a prescient comment to *The New York Times*: "I think they're putting a little lookout tower at the edge of space that the common man can share."

The video stops and MacCallum, seeming satisfied that his maverick credentials are now fully known, awaits questions about his new venture in Tucson, World View Enterprises. World View's concept of operations is deceptively simple: Inflate a massive helium balloon, sling a space-rated capsule under it, and take the nearly two-hour trip to the stratosphere.

A few miles from where we sit in March 2016, construction is underway of a 120,000-square-foot building and large, circular launch pad. From here, World View envisions pods with experiments and paying tourists rising into the stratosphere.

For manned flights, which would carry six customers and two pilots per trip, the pod would ascend at 1,000 feet per minute, cruise at 100,000 feet for a couple of hours, and then detach from the balloon and cruise back to earth under a parafoil.

From the passenger's point of view, the six-hour trip to near-space offers real advantages over more traditional rocket-powered schemes of Virgin Galactic or SpaceX. The difference will be a violent, fast ride versus a leisurely one with a bathroom and a bar. And at $75,000 per seat—cheaper than $250,000 rocket launches offered by Virgin—World View hopes to be a cheaper alternative to get closer to space. (Their trips will fall short of true space by more than forty miles.) The company envisions using high-altitude balloons for climate investigations, to replace communications during disasters, and to predict dangerous weather systems.

A few miles away from where I sit with MacCallum, the construction is already underway. The location will have manufacturing facilities, training rooms, mission control, and an observation deck, as well as a circular balloon launch pad. Taber pulls out some of the blueprints of the property, showing the pad measures seventeen acres.

I drive to the site later. A plain white sign stands on two wooden legs, just outside a chain-link fence that guarding three Port-o-Potties. "Dust Control. World View Permit #8603," the sign says. "Dust complaints? Call Pima County Department of Environmental Quality."

But within three weeks of my sit down with MacCallum, the news comes in. The Goldwater Institute, a conservative think-tank in Phoenix, is suing the county for the Board of Supervisors' decision to fund World View's headquarters and a balloon launching pad. "We're asking the court to put a stop to this deal," Goldwater lawyer Jim Manley tells the local press. "If the county wants to go forward with it, it needs to renegotiate and get a better deal for taxpayers."

At issue is the $14.5 million economic development agreement that the Pima County Board of Supervisors approved with World View to build the spaceport here. Essentially, the county built the headquarters for a rent-paying tenant.

This is how local governments land a business these days. They are almost always at a disadvantage—any business that is offering to buttress the tax base with jobs has a lot of leverage. So the incentives, like tax breaks and new infrastructure, get rolled out. This is standard operating municipal procedure, so much so it's spawned a whole new industry of "site selectors" who work these deals.

And if you don't come willing to bargain, any smart company will walk to another state or town. When you hear politicians say they are "creating private sector jobs" this is usually what they're taking about.

In Tucson, the county is funding the spaceport's construction on a twenty-eight-acre Pima-owned parcel south of Tucson International Airport. World View will pay the county back over a twenty-year lease. Those payments total more than $23.6 million.

This is only a great deal if the company succeeds, and that's what has gotten the Goldwater's hackles up. "The only way that the county can enforce the deal if World View doesn't create these jobs is to cancel it," the local newspaper quotes Manley. "Then World View walks away free and clear and the public is left with the building and a balloon pad that it can't afford and will have a hard time finding a buyer for."

This debate is repeated across the nation, and not just about spaceports. Enticing businesses to relocate has resulted in some classic burns, from empty convention centers to unused sports venues to overgrown golf courses.

Taber, before the lawsuit was filed, said the pushback in Tucson is aimed at the system, not spaceports. "No one comes out and complains about the deals that get cut for a waste management company," he said. "We're just higher profile."

There are other suitors lined up to host spaceports. World View had offers from Florida and Mojave, each offering incentives. Tucson was a natural fit—the founders have been here since their days inside the Biosphere 2 project—but the county still had to fight to keep them.

A media wrangler interrupts my interview with MacCallum. There's a photo shoot scheduled in front of the billboard that the Tucson Metro Chamber just planted in town. The company's founders, members of private space royalty, will be in attendance.

I ask to see the billboard, but what I really want to see is the cast gathered in one spot, a slice of aerospace history. They tell me where the shoot will happen and I zip out there. The billboard has a red pod hovering in space, tethered to an unseen balloon. It says, "Welcome World View."

There at its base stands Jane Poynter, a fellow Biosphere alumnus and Paragon founder, in a floral shirt and purple tank top. Alan Stern, former NASA associate administrator, dresses the part of a former bureaucrat and current mission investigator in a polo shirt and slacks. Mark Kelly, former Space Shuttle pilot and the director of World View's flight operations, wears a skin-tight, long-sleeve shirt. They all wear sunglasses and flash thumbs up.

They are the picture of optimism. They are living sci-fi lives, trailblazing careers that would never existed even five years ago.

By the year's end the company is sending out "save the date" notices for the opening of their $15 million headquarters in February 2017. But the fight over this lighter-than-air spaceport isn't over.

BY 2017, WORLD VIEW IS BRANCHING OUT FROM SPACE TOURISM. They partner with Ball Aerospace, which has been delivering spacecraft equipment since the 1950s, to make high altitude balloons as useful as satellites.

The company calls the concept a Stratollite: a controllable balloon-borne vehicle that can relay information or scan wide areas with any sensors a customer wants. "World View's new remotely managed, uncrewed Stratollite vehicle offers low-cost, long-duration (up to months at a time) persistence over customer-specified areas of interest," the company says.

The arguments for Stratollites make sense. Space launches are expensive and, for the time being, geographically constrained by where they can launch. They are also permanent—you don't bring a satellite back to earth to repair or replace a sensor.

Poynter says the tech will "enable previously unthinkable applications at a fraction of the cost of existing technology." This is only a slight overstatement. There are plenty of times when the US government or a private company would want to build a temporary communications network high in the sky: the National Guard managing a disaster area, scientific researchers operating in remote areas, a Navy task force facing enemy satellite jamming.

The same is true for surveillance satellites. They are powerful but not flexible. Often, they are not where you need them during an emergency, and retasking satellites in orbit, even when feasible, means leaving somewhere else blind. In a national security context, things in orbit are easily spotted and therefore easy to hide from.

Balloons offer a quick, cheap way to watch a large swatch of land for a long amount of time, collecting more data more cheaply than a fleet of airplanes. Existing blimps flown at an altitude of 10,000 feet in the skies over, say, Baltimore could scan an area from upstate New York down to North Carolina's Outer Banks and as far west as Ohio. Imagine the ground a Stratollite can cover from 70,000 feet up.

The potential of this technology is reflected by the company's initial contracts. By February, World View says it has "launched more than fifty Stratollites for commercial customers from sites such as small regional airports, for customers including NASA, Northrop Grumman, and the Department of Defense."

But even as the business model matures beyond just tourism, a simmering fight becomes an existential threat.

The Goldwater Institute filed their lawsuit to request that that county's lease with World View should be "declared unlawful for violating state law regarding appraisals, auctions and rental rates." Just a few weeks before the February grand opening, Judge Catherine Woods grants that summary judgment, voiding the county's agreement with the space balloon company.

"The county is free to renegotiate the lease," Manley, the attorney for the institute, tells a local newspaper. "But only after they appraise the building, hold a public auction, and lease the building to the highest bidder. All of that will protect taxpayers from illegally subsidizing a private business."

It's a blow to the company and the county officials who made the deal happen. The county officials decide to appeal. "While this lawsuit travels through the appellate courts, World View and the county will continue to operate as normal," County Administrator Chuck Huckelberry says in a statement. "World View has moved into the completed facility and begun paying rent."

So despite this cloud, the grand opening of Spaceport Tucson occurs on time. Visitors wander around a balloon-manufacturing table stretching over one-tenth of a mile long, gawk at the one hundred-foot-tall parafoil test tower and take in the slick but empty mission control room.

"World View and Spaceport Tucson are at the forefront of opening an entirely new economy in the stratosphere," Poynter says.

In late February, a few days after the headquarters open, a flight team launches a translucent balloon into the desert air forty-five miles from Tucson. World View and Ball Aerospace are launching a collaborative Stratollite remote sensing mission to test how the sensors and aircraft operate at high altitude. The craft can fly that high safely, but can it be kept under enough control for the sensors to capture high-res images? The balloon silently climbs to 76,900 feet and maneuvers to a single spot. The Ball Aerospace team records panchromatic images of the desert below at approximately five-meter resolution. That's enough for the joint mission to be declared a success, and a new customer base can be courted.

"It paves the way for future flights," says Debra Facktor Lepore, the general manager of Ball's commercial aerospace business unit, "offering higher resolution multi-spectral sensors for applications such as public safety, homeland security, and civic resource mapping and monitoring."

IN THE NINETEENTH CENTURY, FUTURISTS BELIEVED BALLOONS WOULD be the ultimate mode of luxury travel. That explains those mooring stations on top of skyscrapers in New York. Great ships slung under silk balloons would bring travellers on voyages of commerce and exploration. The efficiency of airplanes destroyed that dream, but it's hard not to see its echo in the plans of World View.

The idea that balloons can be used for stratospheric trips is an exciting prospect for budding spaceports. After all, it eliminates most of the problems associated with traditional spaceports. The lack of rockets means an absence of explosives at the facility. Testing equipment in space-like conditions becomes cheaper, and less risky, when using balloons. And you don't even need a runway, just a patch on concrete with some tie-downs.

But there are those who see balloons as an opportunity for true space launches, using them to make that first crucial step off the planet. On March 1, 2017, engineers with the Spanish startup company Zero 2 Infinity boarded a ship and cruised into the Gulf of Cadiz. There,

A Delta IV takes off from Cape Canaveral. If you pair the right launch pad with the right beach, the entire launch is awe-inspiringly close.

Credit: Joe Pappalardo

Massive explosions are always on the minds of spaceport designers. This manifests in bunkers, caches of gas masks, emergency escape systems, and ubiquitous frightening signage.

Credit: Joe Pappalardo

American manned spaceflight rolled to a stop at this spot at Cape Canaveral. The runway now hosts an unmanned military spaceplane but awaits the return of manned vehicles.

Credit: Joe Pappalardo

For more than three decades, Kenneth Hooks has seen Cape Canaveral from his perch inside the nation's most unusual air traffic control tower. The tower controls access to the Shuttle Landing Facility, a 15,000-foot-long runway that hosted shuttle touchdowns until the program ended in 2011.

Credit: Joe Pappalardo

The impressive hardware on display at the Delta IV launch pad at Cape Canaveral.

Credit: Joe Pappalardo

The Cape has never been the same since SpaceX and the Falcon rockets arrived. Seen here in the assembly building on the launchpad, the rocket lays flat until it slides to the pad on rails and then raises to vertical. This makes it easier to work on and fix any problems after a scrubbed launch. The angular shape of the deployable legs are easy to see, tucked tight against the booster.

Credit: SpaceX

A view of an Ariane 5 rocket approaching the launch pad in French Guiana. The crawler painstakingly delivers the upright rocket from its assembly building to the launch pad. It takes hours. The traditional approach may seem out of date but the European firm has an unbelievably reliable satellite launch record.

Credit: Arianespace

The currents that swirl on the islands in French Guiana can be deadly. Zhang Tong, the president of rocket company Great Wall Industry Corp., slipped into the water and drowned at this spot in 1997. His body was never recovered.

Credit: Joe Pappalardo

Mojave Air and Spaceport can be regarded as the birthplace of the private space movement. I took this at dawn after an all night drive.

Credit: Joe Pappalardo

Masten Aerospace is one of the small companies based in Mojave's spaceport. They test engines on tethers, and other times fly and land vehicles under rocket power.

Credit: Joe Pappalardo

Worldview Enterprises plans to launch people to the edge of space in balloons. Here the company founders—an astronaut, two veterans of a biodome experiment, and a former NASA official—gather to see their billboard in Tucson for the first time.

Credit: Joe Pappalardo

The cuisine is often peculiar at parties thrown by the Explorer's Club in New York City. And you never know who you'll wind up next to at the buffet. That's how Jeff Bezos watched me eat a dead roach on a stick.

Credit: Alyson Sheppard

A model of the Lynx spaceplane sits in the Midland-Odessa Airport. The program is on indefinite pause as the company that created it pivots to engine work.

Credit: Joe Pappalardo

A mannequin sits in the Lynx cockpit. The craft would get to orbit under its own power, so the view from here would be epic—if it ever flies.

Credit: Joe Pappalardo

I'm at the spaceport on Wallops Island, standing at the foot of the rocket that launched NASA's LADEE mission to the moon that night.

Credit: Alyson Sheppard

Mid-Atlantic Regional Spaceport Pad 0A during less fortunate circumstances: the October 2014 explosion of an Antares rocket at the Wallops Island spaceport.

Credit: NASA

Spaceport America has a complicated relationship with the taxpayers who helped fund it. Opinions in nearby Truth or Consequences differ.

Credit: Joe Pappalardo

The hangar and passenger processing facility at Spaceport America sits ready for use. Long-promised spaceflights may start in 2018.

Credit: Joe Pappalardo

An illustrated concept of SpaceX's Boca Chica commercial launch site in Brownsville, Texas, released to the media.

Credit: SpaceX

Warning signs mar the beauty of the California coastline at the beach near Vandenberg Air Force Base.

Credit: Joe Pappalardo

A vehicle raises an Intercontinental Ballistic Missile over its silo at Vandenberg Air Force Base. The missile inside will lower into the silo, preparing for a test launch.

Credit: Joe Pappalardo

Pulling an all-nighter with the air force and civilians during a test ICBM launch at Vandenberg Air Force Base.

Credit: Joe Pappalardo

My father-in-law took this photo of a SpaceX booster, recently landed on a barge after a launch, from a Cocoa Beach patio through the eyepiece of a telescope.

Credit: Michael Sheppard

SpaceX launches a reused rocket from Cape Canaveral, a key step to reducing the cost to orbit. An era of cheaper launches is ushering in an age of new spacecraft.

Credit: SpaceX

they lofted a high-altitude balloon with a squat gumdrop and a ring of engine nozzles slung beneath. At sixteen miles above the water, the capsule dropped and the engine flared for a few seconds, the first test of the system that the company hopes will be used to launch small satellites into orbit. By replacing a first stage with a balloon, the company can cut costs drastically and offer launches to universities and small companies, with just two weeks' notice. The company predicts its first commercial launch will be in 2019, and they have letters of intent from launch companies worth $280 million to motivate them.

Such schemes are music to the ears of some spaceports. Places with launch licenses but constrained airspace can use balloons to ferry rockets away from populated areas, over deserts or open water, to launch. Instead of a towering launch pad, with flame trenches and water deluge systems to quiet roaring rocket exhaust, the local spaceport would be quieter than even a small municipal airport.

Other spaceports will be threatened by balloons, however. The advantages of many air-launch schemes (including Waco's) are nullified by using a balloon. For example, rocket companies that want to launch small satellites from mothership airplanes see a good use of their runways. Now, here comes a balloon that can do the same thing but at a fraction of the cost and from a small field.

Stratollite-style balloons are even less welcome competition. Many unmanned spaceplanes can double as temporary communications relays or carry sensors, but those tasks could be replaced by a blimp or balloon. This is a sharp reminder of how a disruptor can become the disrupted in just a few short years.

Such plans may seem like long shots, and in many ways they are. Futurists of any era have a habit of being wrong. But these lighter-than-air aspirations seem a lot closer to reality when speaking with Taber MacCallum or looking at World View's circular launch pad. They're creating a new form of spaceport from a patch of dirt adjacent to an airport. Other surprises are around the corner, as engineers invent new ways to reinvent spaceflight, pushing the entire idea of launch in novel directions.

CHAPTER 10

CAPE REDUX: 2016

SHUTTLES RESTAURANT AND BAR SITS ON THE SIDE OF N. Courtenay Parkway, its weathered sign beckoning to travelers as they drive through Merritt Island. The road connects Cocoa Beach to Cape Canaveral, which is just twelve miles away. It's one of the closest restaurants to Kennedy Space Center, and a favorite of spaceport workers for lunches and happy hours.

Inside, the place is frozen in the mid 1980s. Classic rock serenades the lunchtime diners. The walls are decorated with photos, many signed, of astronauts and shuttle crews. None of the photos are recent: Humans don't leave the planet from Cape Canaveral in 2016 any more than they did in 2012.

The joint sells T-shirts: "Where Astronauts Have Been Lunchin' for Over 30 Years!"

None of the diners today are astronauts. There's a pair of tanned tree service workers in identical t-shirts, guffawing at TV clips of ducks dancing and dashcam car crashes. Retirees sit at another table, picking at sandwiches. There's a table of four doughy men wearing United Launch Alliance lanyards.

Like a parent keeping a bedroom ready for a runaway, Shuttles Restaurant and Bar is waiting for the astronauts to return. Instead of passing through here during training and launch preparations, they are leaving photos at Shep's Bar in Star City, Russia, before boarding Soyuz spacecraft.

The cadence of launches here has been steady as ULA and SpaceX pursue their businesses models, with Bruno's firm hefting government satellites to orbit and Musk's company blasting communications satellites and cargo shipments to the International Space Station. At Shuttles Restaurant, a detailed model of a Falcon rocket stands behind the bar, near the cash register.

But there's a whole new energy around the Cape and an almost palpable sense that astronauts will return. NASA is funding two companies, Boeing and SpaceX, to bring astronauts to the ISS. The agency is also building the Space Launch System, a traditional, government-run program that promises manned missions to other planets, comets or asteroids.

In 2016, SpaceX is putting on a show. The concept of true rocket reusability, as ever at the center of ubiquitous spaceflight, has become a subject of very public engineering. To chronicle the attempts to land a rocket stage for reuse, the company places cameras on the rocket and on an unmanned barge that serves as a landing pad.

The concept of reusable rockets is cool—cool enough that Musk and Bezos have sued over it. The patent office granted the paperwork to Blue Origin in 2014, prompting a legal fight between the billionaires when SpaceX challenged it right away, saying others also developed the idea. In 2015, Blue Origin waved the white flag and SpaceX's landings continue unimpeded.

In April 2016, SpaceX sticks their first landing on the drone ship, after delivering an inflatable space module. They follow that up in May with the touchdown of an empty booster after a launch to geostationary orbit. This feat is even better for SpaceX because that's the kind launch of commercial satellites require, and that market has the sharpest competition over price.

Optimistic timetables are streaming from Hawthorne—SpaceX says it's going to test a Falcon Heavy, the largest rocket ever designed, by the end of 2016. The company says test flights of manned capsules are due in 2017. Their launch schedule becomes stacked with a healthy backlog of commercial and government launches.

And then comes September.

LAUNCHES ARE HIGH-RISK EVENTS. MUCH OF THE DESIGN AND OPERation of a spaceport is meant to keep oxidizers and fuel away from each other. At a Falcon 9 liftoff, kerosene fuel and liquid oxygen ignite, causing an explosion that only well-engineered plumbing can contain and direct.

When these propellants meet unexpectedly, disaster happens.

On September 1, 2016, SpaceX personnel have a Falcon 9 standing on the launch pad at SLC-40, two days before launch. Any rocket launch is a series of countdowns and checklists. One of these engineering rituals is the static fire test, which is designed to test the launch pad as much as it is to test the rocket. Fuel is loaded, lines are purged, cascades of water that dampen sound and vibration are opened. Engineers then pore over the results of the static fire to make sure the launch will be a success.

Static fires are not high-risk operations, typically. The engines are lit only for a few seconds, during the culmination of a static fire test. What's risky about them is that during the test, the valuable rocket has an even-more-holy payload on board.

The September test is no exception. Inside the rocket's nosecone is a satellite worth nearly $200 million, the 5.5 ton AMOS-6 communications satellite. Facebook leased the satellite to provide internet access to Africa, a significant step up for the continent.

The static fire hadn't even begun when a sudden flash blinked from the upper stage. A moment later the rocket erupted in a fireball, a bright orange explosion that started just under the nosecone. The rest is inevitable and ghastly—the power of the fuel to deliver tons of cargo into orbit suddenly unleashed at once, disintegrating the rocket in a split-second, cascading reaction. Like a twenty-three-story fuse, the exploding rocket detonated tanks on the launch pad, sparking a series of booming secondary explosions.

SpaceX will later say that the time between the first signs of an anomaly to loss of data from the rocket is about ninety-three milliseconds, less than one-tenth of a second.

Lt. Col. Greg Lindsey, 45th Mission Support Group commander Detachment 1, gets the call within moments of the detonation. "The call came over the safety net, followed by a call from the Fire Chief. There's been an explosion on Pad 40," he writes later. "We immediately dispatched fire trucks to a staging position and began evacuating all nearby facilities."

SpaceX and the 45th Security Forces Squadron have demarcated an area around the pad dubbed a "Blast Danger Area," the area that is cleared before any hazardous operation begins. The fueling and static fire operation has this designation, so first responders know that no one was hurt or injured in the explosion. "Security was able to immediately account for their officers who formed the cordon of the BDA," Lindsey writes. "Very quickly I was assured that we had 100 percent accountability with no injuries."

Water, not fire, is the immediate concern.

"My civil engineer representative informed me that during the explosion, the deluge system had been damaged and most of the water was being shot up into the air rather than being dispersed across the pad as designed," Lindsey recalls. "The SpaceX rep informed me that while the deluge wasn't functioning optimally it was still helping to suppress the fire somewhat. That was fine except for one thing—our 1.2 million gallon tank was being depleted at a rapid rate and there was no way to refill the tanks fast enough to sustain the output."

Running out of water would cause even more damage, beyond just the burning launch pad. With no water to move, the motors to the tank's pumps burn up. That jeopardizes the deluge systems for other launch pads on the base.

The water has to be shut off, as do any other systems that used gasses and liquids held under high-pressure. "We decided to send in our Initial Response Team to shut down the pumps and turn off one of the high-pressure systems that could be accessed from outside the perimeter of the launch pad."

When NASA responds to a catastrophe, it's all hands on deck. NASA has a fleet of unmanned aerial vehicles, used by a division called Flight Operations and Information Technologies. These UAS operators met with SpaceX reps and work out flight paths to check out the situation before the IRT entered. "It wasn't too long after that first phone call until our incident commander was reviewing footage from the flight over the pad prior to the IRT making its initial entry onto the pad itself," Lindsey says.

The respond teams secure the pumps, saving the spaceport even more damage, and shut down what they can outside the blast zone. But more news comes to the beleaguered responders—pressure is plummeting in the chiller storage units at SLC-41. Before too long, the chiller quits.

When a refrigerator goes out in a house, milk spoils and chicken goes bad. When a spaceport fridge dies, it can take a multimillion dollar spacecraft along with it. In this case, NASA's OSIRIS-Rex is in the chiller, waiting for launch in a week. It's destined for a seven-year trip to return an asteroid sample, so long as it survives the explosion on the launch pad next door.

Frantic ULA and NASA technicians need to break the cordon and get to that chiller and fix it before it is too late. "We were able to coordinate with the KSC Emergency Operations Center for access through their roadblocks," Lindsey says. The NASA mission launches on schedule, a week later.

The Air Force, SpaceX , FAA, and NASA form a team of investigators who start working on what they call a fault tree. Fault Tree Analysis (FTA) is a forgotten benefit of the Cold War, something Bell Labs scientists developed in 1962 to evaluate Minuteman ICBMs. They are useful when there are many variables that could have contributed to an event. The analysts put the explosion at the root of the tree and all the choices and conditions that could have caused it become the branches. This systematic method serves the template for this investigation.

It didn't take too much expertise to see that the explosion started near where the cryogenic liquid oxygen was being loaded. "At the time of the loss, the launch vehicle was vertical and in the process of being fueled for the test," the company says in an early release. "At this time, the data indicates the anomaly originated around the upper stage liquid oxygen tank."

As the billowing smoke clears, there is no question that Slick 40 is closed for business. A support tower has burned to a crisp and the explosion trashed the Transport/Erector vehicle that delivers and raises rockets on the pad. One of the four lightning towers appears singed.

At the pad, the fear of secondary explosions runs very high. Again, the Air Force first responders take a lead role. Explosive ordinance disposal experts, pad safety personnel, and firefighters pace the perimeter, assessing the risks. Air quality monitors sample the environment and binoculars focus on liquid holding tanks, looking for flare-ups. Range rats stare at pressure readings looking for signs of trouble.

"These tasks were tackled by our response teams during almost constant lightning and tornado watches, followed by the pending Tropical Storm Hermine," Lindsey says. "Managing a crisis is one thing; piling on significant weather events only added to the overall complexity."

Something else added to the complexity—spaceport rivalry. Elon Musk starts the speculation train down the tracks with a tweet asking the public to send photos or video of the incident. "Particularly trying to understand the quieter bang sound a few seconds before the fireball goes off," Musk tweets. "May come from rocket or something else."

The idea of sabotage is inherently distasteful to the space community and its veneer of serving the common good of all humanity. The SpaceX investigators don't rule out a deliberate act by a competitor and chase leads that support the theory. "SpaceX had still images from video that appeared to show an odd shadow, then a white spot on the roof of a nearby building leased by ULA," the *Washington Post*'s Christian Davenport reports. "The building, which had been used to refurbish rocket motors known as the SMARF, is just more than a mile away from the launch pad and has a clear line of sight to it."

Could a ULA sniper have taken to the roof with a .50 caliber rifle and scored a hit on the Falcon 9's upper stage? Fuel tanks are tougher than they appear and the smarter moment to take a shot would be during the static fire test itself, when there is noise enough to mask a gunshot.

ULA rebuffs the SpaceX investigator, not allowing him to the roof. ULA called Air Force investigators "who inspected the roof and didn't find anything connecting it to the rocket explosion," the *Post* quotes officials saying.

The conspiracy talk ebbs, and the engineering talk begins. The first thing to examine, as always, is anything different. In December, SpaceX upgraded the Falcon 9 rocket to use even colder liquid oxygen. The colder temperatures are meant to decrease the amount of liquid oxygen lost as it heats and boils off. This extra liquid oxygen is used for cooling stuff like landing rocket stages for later reuse.

Elon Musk described the upgrades on Twitter: "Deep cryo increases density and amplifies rocket performance. First time anyone has gone this low for O2. -340 F in this case." But keeping liquid oxygen supercooled is not easy. "The new supercool liquid oxygen proved to be a challenge during static firing tests," Jason David wrote in a blog for the Planetary Society before the explosion.

In the aftermath or the explosion, SpaceX collects the debris, photographs it, and stores in in a hangar. The data streaming from the rocket begins to tell a tale, helping fill gaps in the fault tree. Over time, and confirmed by examining the debris, the company sees what it calls "a large breach" in the cryogenic helium system of the second–stage liquid oxygen tank.

Ah, the helium bottles again. Recall that after the June 2015 in-flight breakup, SpaceX pinpointed a failure of struts securing tanks inside the rocket as the cause. This incident is seemingly unrelated. But now the integrity of the tanks themselves is being implicated.

SpaceX's investigation stretches to Texas, where engineers test the fuel loading process at McGregor. They load the cryogenically cooled helium into tanks, testing them until failure. Within weeks the company says it can "re-create a pressure tank failure entirely through helium loading conditions. These conditions are mainly affected by the temperature and pressure of the helium being loaded."

In some circles the acronym COPV induces a shudder. It stands for Composite Overwrapped Pressure Vessel, and around Cape Canaveral this refers to the pressurized bottles inside the fuel tank that contain helium. The helium system is used to pressurize the liquid oxygen tank. The name and the method date to the Shuttle program, where the lighter composite shells replaced metal tanks.

But some worried engineers—and NASA employs droves of worried engineers to scrutinize every component—spent months researching the chance that the tanks would rupture inside the larger liquid oxygen tank, causing a chain reaction that could lead to engine failure, or worse. Years of research at NASA research agencies followed.

The hazard was deemed remote. NASA's design ethos is known for its overabundance of caution and its refusal to accept risk, though it wasn't worth trying out a new tank design. As witnessed by the condition of US ICBMs, prudence dictates that it's sometimes safer to do nothing.

NASA did restrict some of the work done on the tanks to preserve the fibers that hold the composite tanks together, a restriction that shuttle engineers asked NASA to remove for the last two flights in 2011. Bowing instead to an abundance of safety, NASA kept the inefficient restrictions in place.

The helium tanks inside the Falcon share the COPV name and concept of operation, but are not built by the same vendor and are used in different operating conditions. So also a direct comparison between SpaceX's treatment of them is not really fair. Or is it? The smart but swashbuckling engineering that enables radical leaps in the private space industry could, this time, have bitten SpaceX in a serious way.

Savvy space watchers know SpaceX had scrutinized the tanks earlier in 2016. They actually switched the order of rockets launching so that they could check out what the company called "bad trends" in a number of helium pressurization system bottles. SpaceX also decided to start building them in-house, rather than go to the Alabama contractor who previously made them.

By early 2017, the verdict is in and made public. The culprit in the disaster is accumulations of liquid oxygen between the COPV's aluminum liner and carbon fiber overwrap that enshrouds it. A slight buckle in the metal liner would provide a pocket for the supercooled oxygen to collect, trapped there when the tank becomes pressurized. One possibility is that the liquid oxygen could have become solid, causing friction to ignite. Damage to the carbon fiber liner then propagates enough of it to burst, starting a chain reaction that annihilates the rocket and its payload.

Now SpaceX has to engage in a struggle like they have not faced before. Instead of charging into 2017 with a full launch manifest and more feats of engineering to win customers and build a spaceport empire, the company has to return to flight and rebuild. The worst damage done is to the trust of customers. While the Air Force said the destruction didn't rule SpaceX out of national security launches contests, several commercial customers decide to go with other launch providers.

The stats remain in SpaceX's favor: As of 2016's end, the firm performed twenty-nine launch attempts of the Falcon 9 in three configurations and delivered its primary payload to orbit in twenty-seven of those times. That's more than 90 percent success—but no one with a valuable payload likes seeing a trend of lost vehicles. After seeing a rookie pitcher lose two games, you have to wonder: Are the stats catching up to the player or has he been hit by a couple of growing-pain performances? Only time and repeated launches will tell.

The explosion validates one large concern that shuttle-era spaceport hands harbor with SpaceX's plans. According to a NASA advisory committee, SpaceX's plan to fuel rockets with the passengers on board, like an airplane, is too dangerous. SpaceX says that their emergency abort system—a bundle of rockets in the capsule that can shoot it to safety during a botched launch—would have saved a crew. But the terms of the commercial crew contract give NASA leverage to slow things down.[10]

---

10. SpaceX seems to be accommodating this NASA concern by designing a crew arm, where astronauts will board the rocket after it's fueled.

Although the launch pad explosion is the obvious cause for the delay, it's really just the highest-profile of SpaceX's obstacles. NASA's inspector general summarized the hurdles ahead, many caused by the company's anti-NASA quest to optimize their performance by changing the way they operate, in a late 2016 report:

> SpaceX has not yet completed all milestones associated with Critical Design Review—a stage in the development process that often reveals shortcomings a contractor must address before it proceeds with full-scale fabrication, assembly, integration, and testing of its capsule. SpaceX officials attributed the delays to capsule design challenges, specifically switching from a design that used a ground-based landing to a water-based landing design in the first year after contract award. This resulted in significant challenges, including complications with vendor components and the effectiveness of the integrated landing system designed to ensure parachutes work and the capsule does not take on excessive water after landing in the ocean. In addition, SpaceX stated it had underestimated the number of interfaces to the weldment and radial bulkheads, which also resulted in design delays.

Elon Musk's goal has been to launch people into space, to first save the United States the $80 million it pays Russia to deliver people to the ISS and, after that, to deliver colonists to Mars. The company and NASA once targeted the initial manned mission for April 2017; in 2016 that date slipped to 2018.

You'd think that the Cape would be downtrodden, seeing their avatar for private space suffer such a setback. But the accident hasn't chilled the mood. In fact, Space Florida's Frank DiBello in 2016 is declaring "a dramatic recovery" since 2011. More facilities are being filled with customers, government spending is going to the government-run Space Launch System rocket, and more plans are being drafted to convert shuttle-era hardware to future operations. More spacecraft are planning to move here, and some of them have seats and windows.

KENNETH HOOKS HAS SEEN NEARLY FOUR DECADES OF CHANGES AT Cape Canaveral from his perch inside the nation's most unique air traffic control tower. The tower controls access to the Shuttle Landing Facility (SLF), a runway that hosted shuttle touchdowns until the program ended in 2011. He can see SLC-40 and SLC-41 from the tower's balcony. He's worked this tower for thirty-seven years, and watched more than a hundred Space Shuttles ascend from one window and land from another, right before his eyes.

Hooks agrees this is the best air traffic control job in the world. In 2016, there's not that much to do—just three hundred operations a month—an amount even modest public airports do in a day. But the things that do land here are world-class and unique, like massive Antonov freighters carrying multimillion dollar communication satellites, the Super Guppy freighter hauling oversize NASA equipment, and C-5s delivering sensitive Department of Defense payloads.

Since 2015 the runway, tower, and other facilities that make up the SLF have been operated by Space Florida, the state's spaceport development authority. The runway's been a symbol of where spaceflight has been. Now it's also a way to highlight the hopes and expectations of the future, from commercial business ventures to the return of human spaceflight to the Cape.

Jimmy Moffitt, the SLF airfield manager, gives a good tour of what he calls "the world's most famous runway." It's a 15,000-foot-long and 300-foot-wide stretch of well-manicured concrete with a surface smoothness confirmed by laser radar mapping. Standing at the centerline, you can just see the subtle curvature of the Earth.

It has two feet of concrete at its the center. There are two wide, asphalt drive lanes on either side of the concrete. There's a plaque commemorating the final shuttle mission near the lip between the concrete and asphalt and another etching in the concrete where the last shuttle flight's wheel stopped.

When the STS-135 mission ended, the fate of Cape Canaveral was uncertain. That included the Space Launch Facility. In 2015, the state of Florida took it over with the intent to make it useful to the emerging private space industry. It remains a private use airfield to preserve the airfield for customers.

Satellite and other payload deliveries are the SLF's trade. Moffitt says he frets until the payloads leave his property. "Once they get to State Road 3, it's not our problem," he says. "But I'll still make calls to my counterparts to make sure they get there alright."

Space Florida's financial support has made Cape Canaveral an alluring spot for the leading private space players. If enough tenants show up, they will need places to park airplanes, store fuel and oxidizers (far apart, please), form companies and process payloads. The state agency is eyeing a four hundred-acre patch of wetland between the SLF and the iconic Vehicle Assembly Building, seen from the tower, to develop these facilities.

"Users are great," says Moffitt. "But we want tenants."

What kind of tenant would use the SLF? Projects at the Cape give Space Florida a reason to keep the runway, hangar, and tower ready. The Air Force has a mysterious spaceplane, the X-37B that launches from Kennedy Space Center on ULA rockets. The experimental mini-shuttle, probably a testbed to prove these UAV spacecraft can replace damaged satellites or launch new ones, can stay in space for years. It lands at Vandenberg for reasons only those with access to classified material can know.

Using this landing strip in California galls Space Florida, which is hoping the X-37B will land where it launches, which does make a lot of sense. The X-37B program is certainly aware of the runway, having tested the spaceplane's landing system during development. Boeing, which runs the X-37B program, also rents a hangar from Space Florida. In 2017, the X-37B landed at SLF, a hopeful sign that one day a fleet of Air Force's space drones could find their way home at Cape Canaveral.

The year 2016 witnessed another marquee firm deciding to set up shop at Kennedy Space Center that opened up the chance of other space-related flights here. OneWeb Satellites has raised $500 million to launch the satellites designed to bring internet access to connect schools in developing countries. The state of Florida kicked in about $20 million in incentive money for the company to set up at Cape Canaveral.

The idea is to crank out communications satellites at an industrial scale, instead of an artisanal one. They will do this in 2017 at a massive manufacturing plant being built at Space Florida's Exploration Park.

It's all good for Frank DiBello. There's nothing better for a spaceport than hosting companies who do work on-site, instead of just using the launch pads or runways and leaving. "We are no longer just launching rockets here, but are building rockets, building spacecraft, and manufacturing hardware to be launched," he told reporters after the OneWeb deal.

But the deal could bring spaceplanes to the SLF. OneWeb has a contract with Virgin Galactic to launch satellites from its LauncherOne system, which tucks a satellite-tipped rocket under the wing of a 747. The company wants its first satellite launched in 2018, but Virgin won't be ready, so OneWeb contracted with Arianespace to launch the first from French Guiana. The good news about this for Space Florida is that, unless OneWorld ships the satellites by barge, the Shuttle Landing Facility runway is the logical place to land the planes and load the cargo for its trip to South America.

The runway could use the action. In late 2016, when I visit, there's only one tenant at the SLF. Moffitt's radio crackles and we have to vacate the concrete to accommodate him.

There's a reedy roar and a moment later a F-104 Starfighter streaks past at around 600 mph. The aircraft consists of a missile-like fuselage, a massive engine, and stubby wings. Icons like Chuck Yeager flew them to break sound-barrier speed records, and NASA outfit them with rocket thrusters and used them to train astronauts. The pilots would blaze to high altitudes and drop, giving them more than a minute of parabolic-induced weightlessness to fire the thrusters and practice rocket-powered, zero-G maneuvers.

The Starfighter overhead sweeps low and suddenly, in an almost cartoon-like defiance of gravity and physics, points its nose to the sky and ascends. That's a smooth word for a powerful, violent motion, but the airplane deserves it because it seems to cleave the air effortlessly. The light of the flaring engine recedes to a pinprick. Watching, it's impossible to not think of a rocket launch.

The F-104s flying at the SLF belong to Starfighter Aerospace, the only private company operating there. The man in the cockpit and behind the company is Rick Svetkoff, a former Navy and commercial pilot who flew F-104s on the air show circuit. He gave that up and devoted himself to buying and flying F-104s for profit.

We meet him at his company's hangar, where a staff of ten tends to the world's biggest collection of working F-104s. Three of the ten, including him, are pilots. "Everything I own are these airplanes," he says. He's an older man with the stocky build of a retired athlete who keeps in good shape. It takes a strong core to withstand high G-forces without passing out.

Right now he rents the Starfighters to test pilot and astronaut equipment. He couldn't do this work anywhere else, except perhaps at Mojave. Although he looked at that spaceport, he stuck with the familiar ground of the SLF and its new, sympathetic state-operated owners. This is not a typical tenant-owner relationship. For example, Space Florida has directly funded experiments with Starfighters. "Besides," he says. "We're Florida boys."

He'll need some political help if his company is ever going to make money flying. The FAA labels Svetkoff's fleet as "experimental" which heavily restricts the business uses of the craft. If the FAA or Congress changes this, he'd be able to pursue other clients. One example: taking experiments and equipment to high altitudes and producing zero-G during parabolic drops. This kind of testing is vital to space scientists and engineers developing components for spacecraft, so much so that he wants them designated as "space support vehicles" and regulated under the more cooperative, or lenient, branches of the FAA that oversee spaceflight. (That would be the Office of Space Transportation, a crucial enabler of the private space industry through its allocation of money and navigation of regulations.)

"We could expand this operation to twenty-five people within six months. We have everything we need," Svetkoff says. "Now we just need to get the regulators off our butts."

Jillianne Pierce, Space Florida's young but capable federal government relations manager, is listening nearby. "We're more than happy to represent and advocate your position," she tells him.

Behind them, a tow vehicle tugs a gleaming Starfighter through the hangar. He turns to watch. Like every dreamer and schemer at this spaceport and so many others, Svetkoff is looking for a vehicle that can deliver him safely to the new world of commercial space.

SO WHEN WILL NEW ASTRONAUT PHOTOS GRACE THE WALLS OF Shuttles Restaurant and Bar? If you asked around in 2012, people would have said it should have happened already. NASA planned for SpaceX and Boeing to deliver people to the ISS in 2015, but now the first flights are scheduled for 2018.

The NASA inspector general lays blame for the delay not on politics but engineering: "While past funding shortfalls have contributed to the delay, technical challenges are now driving schedule slippages."

In Boeing's case, the engineering has proven difficult even for their experienced team. Several former shuttle-era buildings owned by Space Florida now have a new purpose supporting the Starliner, a gumdrop capsule. The program moved quickly at first, not at all the pace of the typical NASA contractor. But that speed proved uncomfortable; in May 2016 Boeing's flight plan slipped eight months. Boeing officials tell the NASA inspector general "in retrospect, its original schedule may have been too ambitious."

The problem is that the spacecraft is too heavy. Well, it has too much mass, since weight is relative to where you are in the universe. In any event, the spacecraft needs slimming, which is usually a component-by-component analysis, a game played in ounces.

Even more worrisome, engineers find that the capsule causes the rocket to be unstable during launch. The Starliner doesn't travel shrouded inside a fairing; it's just plopped on top of the booster stack. The capsule would disrupt the airflow around the rocket during the high-speed flight, so much so that it would cause a crippling loss of control. Boeing's small army of engineers hits the wind tunnels and in October 2016 comes up with an ugly, hexagonal skirt to put around the capsule to smooth the flow. Ugly or not, the simulations show it works.

The news is no better for SpaceX. The Government Accountability Office has a problem with SpaceX's rocket upgrades. Their report claims that there may not be enough time for NASA to review these tweaks before SpaceX's first uncrewed flight test of the Crew Dragon. This prompts a rebuttal from SpaceX president Gwynne Shotwell. "The response to the report this morning was, 'the hell we won't fly before 2019,'" she says.

With all of these warnings, the mood in Florida remains very positive. Nothing lifts the sprits at a spaceport like new construction. If you want to make a Space Florida employee smile, mention the work being done by Blue Origin.

In late 2016, Bezos's company broke ground on a massive rocket manufacturing facility in Exploration Park. The firm and Space Florida are co-building launch infrastructure to launch these rockets, redoing Kennedy Space Center's SLC-36. That SLC is a veteran of the Mariner and Pioneer interplanetary missions, and it will have to be completed redeveloped to host Blue Origin. In early 2016, Bezos gave the first indication of the company's timetable. "We'll probably fly test pilots in 2017," he said. "And if we're successful then I'd imagine putting paying astronauts on in 2018."

Unlike Musk, Bezos plans to fly suborbital capsules before going higher in human-rated orbital rockets and capsules that can get into orbit. He'll vend these services to NASA, satellite owners or anyone else who needs an off-planet lift. It's a direct challenge to Musk and his slow-moving Brownsville spaceport.

The Blue Origin rocket plant construction site is just a skeleton of support beams in December 2016. Trucks, vans, cement mixers, dozers, and cranes move about the site, an ant-like swirl of deliberate but confusing activity. Unlike so much of spaceflight, this is tangible progress on a real-world investment. It represents a new chapter in Kennedy Space Center history. With SpaceX building its own spaceport in Texas, having Blue Origin moving into the neighborhood is a very welcome sight.

At the end of 2016, it's clear that the historic reboot of American spaceflight is underway at the Cape Canaveral spaceport. "It was only five years ago when the Space Coast region was buckled under the loss of almost 9,000 high-wage jobs with the retirement of the Space Shuttle," DiBello sums up. "But Florida is rightfully proud of its leadership role in the dramatic recovery of our space industry."

Not every space scheme, private or governmental, will come to fruition. Aerospace history is littered with good ideas that never flew and bad ones that shouldn't have. But the plethora spacecraft being developed in the United States—and the number of spaceports interested in hosting them—means a greater chance that a new era of space industry has arrived. Like baby turtles on a beach, there's comfort in numbers.

The American aspiration of human space travel, born in the 1960s, was laid to an untimely rest in 2011. If any of KSC's manned programs succeed, this dream will return to where it started, along the eastern shore of Florida.

EPILOGUE: FULL CIRCLE

THE FALCON 9 SITS ON SLC-39A AT KENNEDY SPACE CENTER. It's SpaceX's first launch in Florida since the September explosion, and the pressure is on for a perfect performance.

The slender rocket is sitting on the exact spot where the shuttle launched more than five years ago. The media outlets play up the pad's Apollo heritage—nearly all moonshots left from this SLC—but to me it will always be the site of that last shuttle launch.

Things have come full circle since my first trip to the Cape in 2011. All the shuttles are now museum displays and SpaceX's rocket, once an underdog, is shouldering the heavy burdens.

The final months of 2016 have gone poorly for Elon Musk's company, and 2017 starts with a sour taste as well. The communications firm Inmarsat cancelled with SpaceX and signed a contract with Arianespace to launch a satellite for their broadband-based European Aviation Network. The sting of any lost revenue is painful, but, even worse, this was to be an early customer of the company's Falcon Heavy rocket. That massive craft is what Elon Musk hopes to use to propel himself to Mars. The Air Force is looking at the loss as well as it evaluates ULA and SpaceX for future national security launches.

NASA is also eyeing the Falcon 9's return to flight warily. Sure, there's the near-term need to resupply the International Space Station. But the bigger game is getting humans there, and SpaceX has contracts to use NASA money and missions to pave the way for private manned missions. But launching astronauts is, rightfully, a source of engineering paranoia for NASA.

Losing a pair of payloads has inspired fear for the program, and there are other launch providers moving in to town. Work is progressing elsewhere in Kennedy Space Center on other manned missions, including Boeing's CST-100 Starliner capsule and the traditionally structured Space Launch System, which is aimed at deep space missions. Blue Origin's rocket facility is nearing completion outside the spaceport, another challenger eyeing opportunities in future launches of men and material.

Traditional aerospace can still flex its political muscles. In Washington, DC, legislative aides are writing language to be inserted into the 2017 NASA authorization bill demanding NASA reexamine the Orion spacecraft's "capability to provide an alternative means of delivery of crew and cargo to the ISS, in the event other vehicles, whether commercial vehicles or partner-supplied vehicles, are unable to perform that function." It's a direct slap at private space's ability to deliver a man-rated spacecraft.

The knives seem to be out. SpaceX has something to lose and plenty of enemies looking to carve off pieces of their business. It will take some real-world successes for the company to get back on track.

Credibility is king in space launches, and SpaceX's is on the line. Last year, two weeks after the explosion, SpaceX president Gwynne Shotwell said the company is "anticipating getting back to flight, being down for about three months, so getting back to flight in November. We would launch from the East Coast on Pad 39A in the November timeframe."

But the launch pad preparations make a mockery of the prediction. SpaceX is leasing 39A for manned flights, where the company plans to use the Falcon Heavy. But the September explosion forces a display of adaptability, and the pad is rushed into readiness to loft the Falcon 9.

In the meantime, SpaceX conducts launches from Vandenberg, keeping their schedule from falling into utter ruin. SpaceX launches ten Iridium satellites in a single launch from there in early 2017. It's the first of a planned eight launches from California during the year. It's a necessary sign of progress, a must-win launch in a series of must-wins.

Shotwell is tough under pressure and gives good, strong quotes. I'm jealous when she speaks to Reuters and reasserts that SpaceX's launch pace will reach a cadence of two or three a week. "Shotwell said repairs to the launch pad at Cape Canaveral Air Force Station, which are still underway, should cost 'far less than half' of a new launch pad," writes reporter Irene Klotz, "which she said runs about $100 million."

By February, SpaceX has a legendary KSC launch pad ready for a commercial launch. I compare the 39A with what I saw when the shuttle stood there six years ago. Nothing has flown from here since. It's the same concrete foundation, built for Apollo launches in the 1960s, and the support tower still stands on the pad. But there are obvious changes, like the horizontal integration facility that now stands on the launch pad's perimeter. This building is big enough to accommodate four Falcon 9 rockets, not to mention future Falcon Heavies. It straddles what used to be the track that the massive crawler followed to bring the Shuttle, standing upright, to the launch pad. A water tower has a new paint scheme, now stamped with the SpaceX logo.

Other modifications are subtler. The flame trench and water deluge systems have been redesigned to accommodate the Falcon's engines. New fuel tanks have bloomed on the premises holding RP-1 kerosene.

And savvy space watchers have noticed that NASA and SpaceX have started including the idea of adding a crew access arm to the pad's structure. This small detail should bring a smile to the face of the gang at Shuttles Restaurant, as it heralds the return of manned missions from the Cape.

But it's also a subtle reminder that SpaceX, for all its stubbornness, is adapting its plans to accommodate NASA's protective mindset. The company's original plan was to load astronauts into the rocket, roll them all to the pad, fuel the rocket and then launch. The fueling accident makes people inside and outside of NASA wary that this is an unnecessary risk.

It's possible that a launch pad abort system[11] would have saved the crew's lives had astronauts been aboard when the Falcon detonated, but why take the risk? A crew arm mounted on a tower indicates that the crew will be boarding after fueling, in the familiar Space Shuttle and Apollo tradition.

The countdown is fraught with tension. The weather is good, the crowd anxious, and the ISS crew ready for supplies.

On the radio, mission control officers rattle through checklists, declaring data nominal and systems ready to fly. Then, thirteen seconds before launch, come the dreaded words that no one wants to hear: "Hold, hold, hold."

ALTHOUGH IT'S EASY TO SEE THE REINVENTION OF US SPACEFLIGHT occurring across the country, it's still too early to declare victors. Six years after the last shuttle flight, the space industry is still molting, changing, and adapting. The pace of change is simultaneously staggeringly fast and frustratingly slow.

Being a spaceflight observer, and studying the history of the industry, it's easy to question whether all of this is real. Sure, the spacecraft are real. Engines are tested, airframes flown. Pilots die testing actual hardware. The money spent on them is certainly real. The spaceports that have sprung up to host them have become actual pieces of infrastructure.

But the end goal, the formation of a space industry that is supposed to open up the solar system to our species, always seems to slide

11. The abort system would blast the crew capsule from the rocket to shoot the astronauts away from an explosion, and set the capsule down under parachutes.

away into the future. The breakthrough year always seems to be next year. A cynic would say the government, billionaires, and countless aerospace companies are chasing unicorns, and using as much public money as possible while doing it.

You might think that journalists would be in the vanguard of this cynicism. After all, the more you cover this stuff and the closer you get, the more evidence you accrue that this space industry is a scam run by a cabal of corporate welfare recipients and uber-nerds.

Visiting American spaceports is the cure for this cynical malaise. I find that the closer I am to spaceports, the more apt I am to believe in this new age of spaceflight. Not that I think everything I hear will come true—not every spacecraft will fly anywhere. In fact, most won't. Spaceport designations will go unused, vehicles will be mothballed, contracts will be cancelled, and companies will dissolve. Any evolution includes extinctions.

But it's the volume of activity that gives me some faith. Only a couple of space projects need to survive to revolutionize spaceflight in America. A single successful air-launch program could change the small satellite launch market. A private company delivering manned capsules into orbit would reinvent the industry. Any company that can bring a paying customer to the edge of space would create a new category of tourism.

For each of these dreams, there are several players working on real-world projects. This heartens me. I don't want any of the baby turtles to be eaten before reaching the water, but when it happens there at least will be another one racing to the surf.

So far, although beset with accidents and disrupted timelines, most of the big name spacecraft projects remain viable, meaning hope still flickers in the spreadsheets, launch pads and runways of America's spaceports.

This is evident in a 2017 examination of the vertical launch industry. Jeff Bezos is emerging as a player at Cape Canaveral, though his impact on the industry still unknown. His firm has not launched anything on a real mission, but his canny ties to traditional aerospace and willingness to build a rocket factory at Cape Canaveral are tangible signs of Blue Origin's seriousness.

Orbital, now called Orbital ATK after a merger with Alliant Techsystems, is planning ahead. With the Antares rocket back in the game, the company has a busy launch schedule in 2017. A slate of improvements to the rocket and capsule (between 2017 and 2019) will enable them to carry more weight. The plan demonstrates the company's positioning to compete for NASA science missions and commercial contracts. Deliveries to the ISS continue, and although the company swallowed its pride and launched its Cygnus capsule on an Atlas V, the company and Wallops Island are still very much in business.

ULA's involvement in Orbital's launch business is another sign of Tory Bruno's new way of doing business. The company is getting into fighting shape, shedding hundreds of jobs and designing its Vulcan rocket line to replace the Atlas V and Delta IV with a cheaper alternative. In a Bruno-esque display of marketing savvy, ULA created a website where customers, space press and interested public can "create their own rocket" to showcase their launch options. The monopoly that everyone loves to hate is still very much in business, and ready to fight like Templars for market share.

The horizontal launch/spaceplane business is also moving ahead, albeit slowly. The X-37B, that secretive unmanned spaceplane operated by the Air Force, is in operation and is using the long SLF runway at Cape Canaveral to come back to Earth. Dream Chaser has been knocked out of all NASA contract competitions for now, edged out by tried-and-true capsules. It has an uncertain future past an upcoming test flight, but rumors have started to circulate that it could be used to maintain the Hubble Space Telescope if its replacement doesn't launch on time.

None of the manned spaceplane projects envisioned to roost in spaceports in Texas, New Mexico, or Florida are flying missions in 2017. Virgin Galactic is still moving ahead, with SpaceShipTwo making its first free flight in December 2016. This would be the first of probably more than a dozen free flights, but its clear that the fatal accident hasn't derailed the effort. The company's pursuit of air launched rockets to loft small satellites yielded launch contracts from NASA and OneWeb, and it could take the skies as early as 2018.

Another player has emerged from the shadows in Mojave. Stratolaunch, the world's largest plane, eased its way out of a hangar in early 2017. The brainchild of billionaire Paul Allen and maverick designer Burt Rutan, the enormous craft has an incredible presence but an uncertain future. They can't seem to find a rocket that suits the plane, even after commissioning several companies. The market has also shifted. Envisioned years ago as a way to launch mid-sized satellites on demand, the market has moved to smaller payloads. Why use the world's largest plane to launch a small rocket? Still, it's a welcome sign that the airplane has left the hangar, with the promise of flight tests to come.

Astronauts are supposed to board the SpaceX and Boeing capsules for first flights in 2018. This milestone will be historic—the US return to manned spaceflight after a seven year hiatus. Both companies in early 2017 claim they will make this date. They'd better—it takes three years to negotiate a launch contract with Russia. If Boeing and SpaceX don't start flying, and soon, creating a multimillion-dollar backup plan overseas could neuter this part of the young private space effort.

These are the stakes when SpaceX's launch is scrubbed in February 2017. But it's just a delay, and ice-veined space watchers know the company is being careful. Those who don't know this are informed by Musk via Twitter: "All systems go, except the movement trace of an upper stage engine steering hydraulic piston was slightly odd. Standing down to investigate." Later he adds: "BTW, 99% likely to be fine but that 1% isn't worth rolling the dice. Better to wait a day."

Spoken like a true, cautious launch provider.

THE SAME CHARACTERS ASSEMBLE AT KENNEDY SPACE CENTER THE morning after the SpaceX scrub and perform the same drama. Inside the Dragon capsule on top of the rocket is an experiment to grow stem cells in microgravity and another test to grow antibodies in space. There is also a load of supplies, including food. "Looks like I'll have to wait a day to get my cheese," quips one French astronaut onboard.

This will be the tenth load of cargo delivered to the space station—if SpaceX can get it off the ground.

The countdown begins anew. This time, there is a release of tension as the countdown crosses the twelve-second mark. The last ten seconds seem like a formality and, indeed, the rocket's nine engines flare and the Falcon leaps into the air. At fifty miles high, the first-stage booster separates. The payload continues into space, riding on the second stage.

The first stage coasts to about a hundred miles in altitude, then begins to descend. Three of nine engines reignite. The rocket has its own guidance-control system, which gimbals the Merlin 1D engines to rotate the rocket between 120 and 180 degrees. This aims the rocket toward a designated landing pad on the ground. The booster is traveling at nearly 3,000 miles per hour.

The center engine ignites to slow the descent and gimbals itself to help the booster become fully vertical. Four grid fins extend—they look like thick fly swatters—to stabilize and further slow the cylinder. Each paddle moves independently to control roll, pitch, and yaw. The rocket's speed drops from 3,000 miles per hour to about 560.

The landing burn begins. The engines initiate a final burn to slow the craft to about five miles per hour. Four carbon-fiber-and-aluminum-honeycomb landing legs unfold, powered by compressed helium. The stage touches down, a ring of dust and dirt swirling under the engines' lick of flame. It looks like a takeoff in reverse as the rocket settles down and halts peacefully.

And just like that, the future returns. SpaceX has regained momentum with a successful launch and booster recovery. Their customers will never forget the $200 million payload ruined in November, but the intrepid launch company is operating as if they are only as good as their last launch. And this one appears to be an unqualified success—an improved rocket, a marquee client, a reused rocket stage, and, most of all, a new launch pad.

The empty rocket stage is now ready to be cleaned up and prepped for reuse. SpaceX says they can lower their launch price by 30 percent if they can reuse these boosters. This feat of engineering has never been done before, and despite Tory Bruno's skepticism, promises to open spaceflight to more people. The few SpaceX employees authorized to speak to the media have a standard question: How expensive would airplane travel be if they threw the plane away at the end of the trip? They are still promising the democratization of space travel, achieved by lowering launch costs. For all the drama, setbacks and come-from-behind victories, this ethos has remained unchanged since Musk founded the company.

This dream takes another significant step forward in March 2017 when SpaceX launches a communications satellite into orbit using stages of a rocket that had already launched and landed during an April 2016 mission. This is a historic first. The satellite reaches orbit successfully and the booster lands on the SpaceX drone ships floating in the Atlantic Ocean. It's a perfect day for SpaceX.

SpaceX CEO Elon Musk appears on the company's live stream to deliver a message. "It means you can fly and re-fly an orbital class booster, which is the most expensive part of the rocket," he says. "This is going to be, ultimately, a huge revolution in spaceflight."

By June, the gang from Hawthorne is at it again. This time, it's the spacecraft itself that is to be reused, the first to lift from Cape Canaveral since the Space Shuttle retired. And it's even rising from the shuttles' former launch pad, 39A.

It's easy to tell the difference between a Falcon rocket on a space station resupply run versus one prepared for a satellite launch. The size is the same, but the tip looks much different. Satellites need extra protection from the aerodynamic loads and other stresses of launch, so there's a fat fairing on top for those missions. But Dragon is a spacecraft, dammit, and needs no such coddling. It's an aerodynamic gumdrop, suitable for high-speed flight and fiery reentry. So the tip is smooth and slender.

This capsule has seen space before. In 2014, it delivered thousands of pounds of supplies to the ISS and returned to an ocean splashdown. For years NASA insisted on using only new capsules for these missions, but SpaceX wore them down by pitching detailed plans on how to re-certify them for flight. The biggest challenge isn't the violent launch, hostile space environment, or fiery reentry, but the effect of saltwater on the craft. That's why the next iteration of the Dragon capsule will land under rocket power, like the returning boosters, and set down on land instead of water.

"That's a perfect step in the pathway to crew," Benjamin Reed, director of commercial crew mission management at SpaceX, told the International Symposium for Personal and Commercial Spaceflight in 2016. "We'll get very comfortable with doing propulsive landings with cargo first, and then with crew."

It's one thing to hear company officials say they are charging ahead with these plans, but it's another to see the evidence blasting off from a launch pad and zipping through space.

The June 2017 launch is familiar, almost routine. SpaceX has festooned their hardware with video cameras, which is useful for engineers but also for public relations. The Friday launch, in true Cape Canaveral tradition, is delayed a day for lightning.

SpaceX's video feed is split on the screens, to capture all the action. There's a lot to watch. The rocket stages separate, showing vistas of the planet below. Nozzles on the empty booster flare as the empty tank orients itself for a landing. This time the empty engine sets down on a pinpoint landing on Air Force-owned land instead of the drone ship. As cool as those water landings are to behold, it's much easier and cheaper to just land an empty stage close to where it will be refurbished and refilled.

At the same time as this landing occurs, the second stage is flaring, to another, higher view of the earth's curvature. Trips to the space station are timed nearly to the second, a careful choreography of orbital insertion and a slow approach, engine burn after engine burn, to catch up to the structure. NASA describes this a stair-step to the station.

Hundreds of staff at SpaceX headquarters in California cheer each milestone. These employees gather to watch every launch. Over the years I've met lots of SpaceXers, from former astronauts to engineers to range rats. Many have an aspiring rock star, geek celebrity air about them, like the guys at the Magic table with the coolest cards. They're not allowed to speak to press, but they've proven to be great fun off the record, when their ambition and expertise combine in mind-bending conversations.

One of my favorite interactions comes in 2017, when I tag along with some Space Florida people to a National Space Club happy hour at a bar in Port Canaveral. It turns out that the event's date changed, but the organizers didn't spread the word. Nevertheless, a table of SpaceX employees are snarfing bar food and quaffing beers at a table, and invite us to sit. There's a one to three ratio of female to male, and the ages skew young. No one at the table is more than thirty-six years old. The conversation drifts from the local scene to far-flung test spaceports—one older employee remembers launches from Kwajalein—but settles on the future. Always the future. They scoff at competitors, their lack of vision, and aversion to risk. With smug, plausible conviction, they believe their company will not only get humans to Mars, but start a new chapter in human civilization.

As the Dragon capsule makes its way into orbit, I wonder where the guys I met in Florida are at this moment. I imagine the telemetry guys are on the range somewhere, in a windowless room staring at a screen. The propulsion folks may be working the launch inside mission control in California, examining pressure sensor data. The rest may just watch from the crowd in California.

The SpaceX culture is developing as brash, capable, and hard-charging. (Even an internship listing starts by asking if candidates are "undaunted by the impossible.") Their braggadocio can be off-putting, but such pride and dedication is mandatory. Breaking the mold is never easy, especially when extreme engineering, government contracting, and orbital dynamics are involved.

That attitude is catching. Spaceflight will never be as cool as, say, being in a band, but the breakneck attitude is a far cry from what I saw in 2011, when mopey graybeards stalked the spaceport just waiting for their programs to die. It is a time for iconoclasts, for rebels with hubris, and for unbridled ambition.

I do wonder about the young people who are only now entering or graduating from universities. Are these interns, fledgling engineers, and newly minted range rats watching every launch, as I do, or are these moments coming so regularly that they accept them as the norm? I remember James Pura, director of the Space Frontier Foundation, once telling me that reusable rockets "will lead to space travel being just as easy and inexpensive as traveling from New York to London is today." He left out one part: These events will also be routine, unlauded, and as underappreciated as an airline flight to Boise.

The next generation of the space industry workforce may not realize that the vibrant landscape they are enjoying could have been stymied. I consider it nothing short of a miracle that the private space plan that has reinvented American spaceflight has survived three administrations in Washington, DC. And not just any three: We're talking about the wildly different regimes of George Bush, Barack Obama, and Donald Trump.

SpaceX is the face of this miracle. While others have worked on test launches and designs, they have been igniting rockets and completing missions in orbit. Elon Musk has paved the way for the ambitious new players and inspired the complacent launchers across the world to reinvent. SpaceX could fold tomorrow and still be the most important launch provider in twenty-first century history. Given this, a little grandiosity at the bar can be expected.

KENNEDY SPACE CENTER IS NO LONGER JUST A GOVERNMENT-RUN facility. It's a spaceport, open for business for any company that can convince people to pay for a rocket launch. It remains home to the ambitious, the brilliant, the brave, and the stubbornly persistent. Cape Canaveral is at the center of the ongoing renaissance, and it now represents the future.

KSC's rebirth is mirrored across the country as more and more launch centers are founded, built, and designated. More airports are welcoming space companies, more runways are being lengthened to accept spacecraft, and more aerospace jobs are aimed at making money above 65,000 feet. With this future comes backroom politics, raging egos, raids on public money, an embrace of danger, and the heartache when the engineering goes wrong.

But then I imagine this nation without its spaceports, of barren desert towns that abandoned their aerospace heritage and the uniquely American drive that goes with it. I wince thinking about the sweet expanse of American airspace going unused by entrepreneurs and the airports that closed instead of rolling the dice by embracing a new form of transport. I envision American manned spaceflight without private space involvement—a sluggish NASA program with no steady political vision or even a destination. And I can almost see a generation of young engineers turning away from aerospace, discouraged by a lack of real-world projects.

I'll take the freewheeling science fiction future, warts and all.

Melville's description of the ocean comes to mind again. The "harborless immensity" of space now has more American ports of call than ever. From coast to coast, space has never been closer.

SPACE LINGO

One thing I've never understood in books is a glossary. They only appear to be helpful, but they're not so useful tucked at the end of a book. If there are terms you need to know, the author really should explain them on the page.

So don't call this a glossary; instead, this is collection of space lingo, terms, and concepts that didn't make it onto the page in the main text, but are still useful to know. You're done with the book, loyal reader, and since you've come this far it's safe to assume you'll be watching as the reinvention of spaceflight continues. I hope these definitions will help beef up your command of space jargon and prepare for the journeys ahead.

Escape Velocity

Toss a ball into the air and it will rise, pause, and then drop. A rocket scientist would look at that and say, the ball's speed was slowed by Earth's gravity, which pulled it back to the ground. To get the ball off this planet means reaching a speed that can overwhelm Earth's pull. Ask anyone, including Google, how fast that is and the response will come back: about 7 miles per second, or 25,000 miles per hour. (This figure doesn't take into account aerodynamic drag and air friction, so the speed actually must be a little higher.)

Understanding this challenge is useful when contemplating space rockets. It's why a rocket's weight is more than 80 percent fuel, a massive amount needed to achieve this kind of sustained thrust. It also explains why rocket flights start so slowly and ramp up their speed at higher altitudes. The air is thicker at lower altitudes, and that resistance makes it inefficient to try and go fast. That fuel is better used at higher altitudes, since the rocket can more easily reach higher velocities where the air is thin. A steadily accelerating flight profile also keeps payloads and astronauts from getting broken and squished during a launch that would go from zero to 20,000 mph straight off the pad.

Orbital Depots

Weight is the most vital variable in a space launch. The heavier the payload, the more fuel you need to get into space. But if you burn all your fuel getting to space, what will you use to zip around once you're up there? Adding extra fuel will, perversely, require more fuel to carry it. Breaking this frustrating spiral will become an issue as space exploration plans become more ambitious. When the discussion turns to manned exploration, the supply issues become even more problematic. One oft-discussed solution is the "orbital depot," a prepositioned cache of rocket fuel and water launched into space, waiting for a spacecraft to swing by and top off their tanks. These could be right here in Earth's orbit or positioned farther out in space. (The difference being, a depot in orbit would need thrusters and fuel to maintain its orbit while one far enough from the planet wouldn't.)

Those who agitate for space-based industry wonder why we should launch supplies from the planet at all when supplies can be more easily harvested from asteroid and the Moon. For example, water ice mined on the Moon can be used to stock space depots. Delivering it from the Moon is far easier since the energy (re: rocket fuel) needed to achieve escape velocity there is far less than an Earth-launched rocket. This goes back to escape velocity—the Moon has less gravity and no atmosphere to cause drag. That's a cool sci-fi solution, but near-term spacefarers will likely launch and rendezvous with a previously launched cargo segment of their spacecraft before departing for their cosmic destination.

Space Industry

That first step off the planet, as we've seen, is the hardest and most expensive. So what can we actually do up there to make money, besides beaming text messages and porn gifs to cell phones? There are companies looking into asteroid mining, hoping to bring valuable resources back to the planet on the cheap. Others see the benefits in medical research, mostly stemming from the way the absence of gravity makes it easier to tamper with cellular walls while doing genetic engineering.

These are exceptions. There are a lot more schemes to make money servicing space industries than there are actual budding space industries. In Gold Rush terms, there are a lot more people selling gold pans and picks than there are prospectors. These services include creating supply depots, manufacturing space-made solar panels, mining water from lunar craters, setting up space navigation aids, and establishing trucking services between solar system waypoints. Who will be out there buying the new solar panels and tanks of asteroid water is less clear. For the near future, anyway, the commercial money in spaceflight will be made with the delivery and operation of satellites in Earth's orbit as deep space industry limps along.

In a more practical predication, that means the competition between rocket companies at spaceports for government launches will remain fierce. Hopefully that will drive innovation and give new spacecraft a chance to earn a place in any future off-planet economy.

Luxembourg

Talk about the mouse that roars. This landlocked country is wedged between Germany, Belgium, and France, with a population of about 600,000. But this nation's impact on the solar system may make it the most important flag waving in space. A new Luxembourg law that came into effect in 2017 recognizes the legal ownership of resources obtained by private companies. That means an asteroid mining firm like Planetary Resources can conduct their business without worrying whether or not a nation will back them. The response from the industry has been clear—a handful of space companies, including US-based Planetary Resources, have already agreed to move there.

Lawmakers in Luxembourg say their law is fully compliant with the Outer Space Treaty of 1967, ratified by 107 countries, which doesn't allow any nation to claim a celestial body. Less helpfully to space miners, the treaty also claims exploration of space needs to be done "for the common heritage of mankind." Undaunted, Luxembourg is issuing licenses to space companies. The nation has done this before and benefitted: In 1985 they helped create the satellite company SES, breaking with European tradition. In 2017, SES's forty-two sats generated two billion dollars a year. So if you see fleets of spacecraft launching with three red, white, and blue horizontal bars stenciled on their hulls in the future, you'll know why.

LOC

The space story of 2018, hopefully, will be the return to manned spaceflight in the United States. The standard line you hear from NASA is that manned flights must be safeguarded against all calamity, that there's no room for error. A fatality during a launch could derail the entire effort. This is all true, but it looks different when seen through the lens of NASA bureaucracy. All of this heated debate and worry is encapsulated in a single term: Loss of Crew, or LOC. This is the acronym that can doom the Commercial Crew program to a downward spiral of delays. The LOC is a measure of the probability of death or serious injury to passengers inside a spacecraft. They determine this by running tons of simulations and scenarios, and examining the ways the rocket and capsule design mitigate those risks. NASA's contract with Boeing and SpaceX require the LOC should be one bad thing per 270 missions. This is the first program to ever demand that meeting the LOC be part of the qualification to fly. It can also be noted with quiet snark that the Russian Soyuz that currently ferries astronauts to and from the space station has never faced any NASA qualifications, not to mention this one. The Space Shuttle's LOC, over its career, was about 1 in 90.

In any event, the GAO has issued some warnings in 2017 that achieving the LOC might be a problem. For example, the companies at first didn't factor in a bird strike during liftoff or splashdown. With some scenario building and creative thinking, these risks can be reduced (in case of a birdstrike, for example, using ISS exterior cameras to make sure the capsule wasn't damaged.) Other risks are focused on the capsule's ability to take a hit from space debris or being holed by a micrometeorite.

Safety is paramount, everyone can agree to that. But it seems odd that all the engine development, test flights, material science, and crew training could be put aside for a rather arbitrary ratio based on simulations and role-playing. Then again, I'm not the one riding in the capsules, so that's easy for me to say.

MaxQ

There's a ton of jargon you hear from mission control during a launch. Listing them all would require a dedicated glossary. But there is one phrase that means more than any other to aeronautical engineers: "Approaching MaxQ." This stands for "maximum dynamic pressure" and it's a scary moment. It's the part of a launch when the rocket is under the highest amount of aerodynamic stress it will have to endure during its flight. Understanding MaxQ is pretty easy. The two things that influence it are air density and speed of the rocket. The thicker the air—in other words, the lower the altitude—the higher the aerodynamic stress becomes. The faster it goes, the more stress the vehicle must handle. Luckily, these two factors are going in opposite directions. The air is thinner the higher the rocket flies and it's going faster, too. Plot these factors on a chart and the place they intersect is MaxQ. One popular way to avoid stressing out these rockets is to cut the amount of fuel going to a rocket, throttling the speed down until the craft gets a little higher and can go faster without risk.

Taikonaut

For all the lofty talk about space being the shared dominion of all mankind, the truth is that space travel and geopolitics are pretty closely tied. As with many things, the proof is in the language. Spacefarers are called different things by different countries. Americans and Europeans have astronauts. Russians have cosmonauts. India's space program coined the term vyomanauts, based on the Sanskrit word for "space." And now China has taikonauts.

The Chinese space program is intense, and has made great strides since its inception in 2003. Within ten years they landed a probe on the moon. There are new launch vehicles being lofted, unmanned space stations put in orbit, and bigger rockets planned that are intended to help get taikonauts into space and onto the surface of the Moon. The nation hopes to have a manned space station up and running by 2022.

The reaction to it greatly depends on your view of the nation. Those who focus on its military aggression and human rights violations see a potential adversary to the west seizing the high ground of space. Those who see opportunity want a shared space program to be a path to maintain peace and cooperation. Either way, there will be a lot of interesting developments in China, as taikonauts become bigger players in the space races of the twenty-first century.

Space-based Surveillance

I love the way satellites are portrayed in movies and TV shows. The sats are always where you want them, capturing images with impossible resolutions that can be "enhanced" to reveal clues. The reality is actually more exciting—smarter satellites and constellations of small sats are making it easier for firms to offer ubiquitous satellite coverage. Want to track seagoing ships? We have a constellation for you. Does your university need to study ice coverage in Greenland? We got you covered, for a fee. So "space-based surveillance" can mean "satellites watching things on Earth."

But there's another use for the term. When someone in a military uniform talks about it, that person is referring to a ten-satellite constellation operated by the Pentagon that tracks space objects in orbit. This means watching for space junk that could pose a threat to hardware and people in orbit. As other nations get more capable space programs, this also means tracking suspicious objects that could be observing, tampering or threatening to destroy American communication or spy sats. Both of these trends, the commercial and the military, are only going to grow in the years ahead.

Mega-constellations

Satellites are getting smaller but plans for them are getting more ambitious. Mass production of these smaller satellites will change how earthlings use space, and require a new way of organizing traffic in orbit. There are big name companies backed with serious money who are preparing to launch large numbers of small sats. SpaceX's plan alone calls for thousands. This makes sense since the loss of one won't compromise the network and upgrades are easier to implement. It also enables smaller rockets to launch the sats, making it cheaper to set up a rugged space network.

However, these plans are meeting pushback. Some feel these small sats will make the space junk problem in orbit even worse. Others have a problem with the rule changes needed to make these schemes work. International regulations enable sat operators to operate on a first come, first served basis in orbit, i.e. you get priority use of the spectrum if you own the hardware. The FCC rules say that sats in non-geostationary orbits must share the available spectrum when they're close to each other. The FCC seems to be giving ground to the international rules, meaning new constellations will have to avoid spectrum overlap with other sats. How this plays out will determine who wins the mega-constellation battle, with big implications for places like Cape Canaveral which hosts constellation maker OneWeb and its rival, SpaceX.

SELECTED SOURCES

50 Years of NASA History

NASA, Steven J. Dick, NASA Chief Historian

NASA has a robust archive of its history, which provides a welcome window into the agency's early years. Much of it is surprisingly candid and useful. The article cited below ran as part of a series of articles celebrating the agency's fiftieth birthday. (https://www.nasa.gov/50th/50th_magazine/historyLetter.html)

Pancho Barnes' Happy Bottom Flying Club

Lay of the Land Newsletter, Center for Land Use Interpretation, Fall 2000

The non-profit CLUI is obsessed with what humans do with the land—not in an environmentalist way, necessarily, but by cataloguing what humans have done. To explore its online land use database is to risk a whirlpool-sized time suck. Their approach is often artistic and this keen sense of the peculiar informs their write-ups of fascinating places, like the Happy Bottom. (http://clui.org/newsletter/fall-2000/pancho-barnes-happy-bottom-flying-club)

R-2508 Complex Users Handbook

Edwards Air Force Base, January 2017

I'm always amazed at the information that you can find about government property. In the case of this amazing piece of airspace, there are enough people who take advantage of it that it has this publicly-available users handbook, which even includes this handy map, below. (http://www.edwards.af.mil/Home/R-2508/)

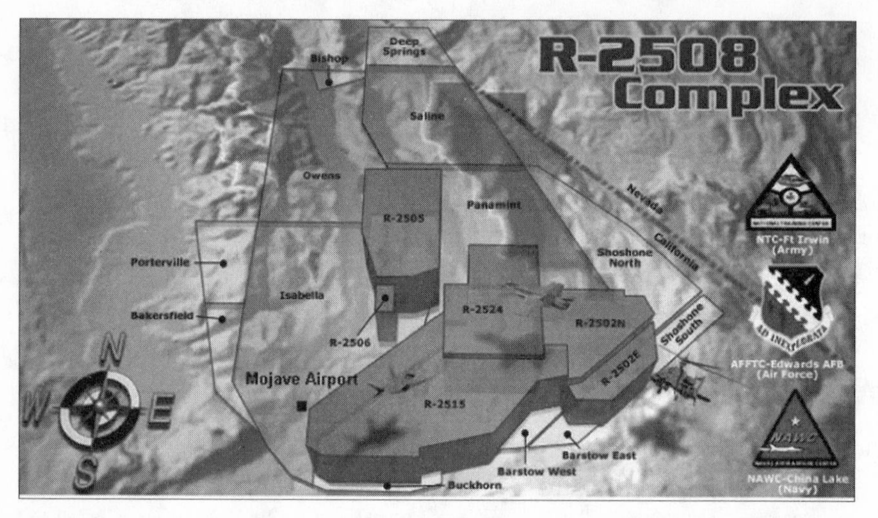

Good American airspace is an oft-overlook national asset.

Photo Credit: USAF

Commercial Space Transportation Overview

Office of Space Transportation, May 31, 2017

The United States government runs on PowerPoint slides. To read one is to tap into the bureaucratic mind; organizational charts and jargon are mandatory. But like many of these things, OST's overviews are nice snapshots of the commercial space industry from the FAA's vantage. (https://www.faa.gov/about/office_org/headquarters_offices/ato/service_units/systemops/ato_intl/documents/cross_polar/CPWG23/CPWG23_Brf_Commercial_Space_Transportation_Intro.pdf)

NASA's Response to Orbital's October 2014 Launch Failure

Office of the Inspector General, 2015

Not many government IG reports let the veil slip on the authors' frustration and anger. This document is the exception: a barnburner cloaked as a dry recitation of an investigation into insurance and Congressional budgeting. (https://oig.nasa.gov/audits/reports/FY15/IG-15-023.pdf)

"Wallops Island - 60 Years of Exploration"

NASA, September 2005

Here is another good example of NASA's thorough cataloguing of their facilities. Wallops has never been the most popular or A-list launch facility in the world, so having information about its founding is a relief for spaceport freaks like me. (https://www.nasa.gov/vision/earth/everydaylife/wallops_60th.html)

Final Supplemental Environmental Assessment for the Blue Origin West Texas Lunch Site

Federal Aviation Administration, February 2014

Environmental impact statements are like tall ladders. They enable nosy people on the outside to see just a little of what's going

Blue Origin's unmarked main gate outside Van Horn, Texas.

Photo Credit: Joe Pappalardo

on inside the fence line of private rocket companies. Otherwise, a view like the one pictured below is the only glimpse Jeff Bezos will provide of what's going on inside his Texas launch facility. (https://www.faa.gov/about/office_org/headquarters_offices/ast/media/Blue_Origin_Supplemental_EA_and_FONSI.pdf)

Office of Commercial Space Transportation; Notice of Availability of the Final Environmental Impact Statement for the SpaceX Texas Launch Site

Federal Aviation Administration, Department of Transportation, June 2014

I told you I love these things. Here's a blurry peek at the SpaceX engine facility from a nearby road. The engine test stands are easy to see, but keen-eyed readers may pick out the white gumdrop space capsule. (https://www.federalregister.gov/documents /2014/06/06/2014-12985/office-of-commercial-space-transportation-notice-of-availability-of-the-final-environmental-impact)

The view from the perimeter of SpaceX's operation in McGregor, Texas.

Photo Credit: Joe Pappalardo

List of Office of Commercial Space Experimental Permits

Federal Aviation Administration

The FAA seems downright proud of itself for doing what it's supposed to do: issue licenses to private companies to build and operate spacecraft. Their website has an updated list of experimental permits, a window on the actual commercial space work being done across the nation. (https://www.faa.gov/about/office_org/headquarters_offices/ast/environmental/review/permits/)

Emergency Management: A behind the scenes look on the Eastern Range

Lt. Col. Greg Lindsey, September 2016

I'm indebted to this dramatic but hardly sensational write-up of the harrowing explosion at Kennedy Space Center. Firefighters are brave, but spaceport firefighters have a slew of additional troubles to contend with, like toxic chemicals and volatile explosives. I've seen them firsthand put out fires (Vandenberg) and treat heat stroke victims (Spaceport America.) This is my hat

This is the sight that greeted firefighters at Kennedy Space Center in 2016.

Photo Credit: NASA

tip to them. (http://www.patrick.af.mil/News/Commentaries/Display/Article/938481/emergency-management-a-behind-the-scenes-look-on-the-eastern-range/)

"Anomaly Updates"

SpaceX release, January 2017

SpaceX's media approach is both revolutionary and frustrating. For a company that competitors say exists only because of its PR machine, they have a very lean media operation. It's hard to claim and keep their attention—part of you knows that it's because they are too busy to idly banter with mere reporters. (Having said that, Elon Musk's team has always made time for *Popular Mechanics*, even when I was being a pain in the ass.)

But the real appreciation I have for their media operation came when they live streamed the attempts at rocket booster returns. Each failed landing became a lesson in leading-edge engineering—no test is a failure if you learn something. And when they stuck the landings, we all felt the joy of the experience. I know young engineers who joined rocket clubs based on the experience. (http://www.spacex.com/news/2016/09/01/anomaly-updates)

Commercial Space Launch Amendments Act of 2004

US Congress

This is the one that started it all. It's a revolutionary law that enabled the industry that this book chronicles. And its impact will only grow as time goes on. "Private industry has begun to develop commercial launch vehicles capable of carrying human beings into space," it says. "The regulatory standards governing human spaceflight must evolve as the industry matures so that regulations neither stifle technology development nor expose spaceflight participants to avoidable risks." Good stuff. (https://www .faa.gov/about/office_org/headquarters_offices/ast/.../PL108-492.pdf)

ACKNOWLEDGMENTS

THIS BOOK WOULD NOT BE POSSIBLE IF NOT FOR A LONG LIST OF helpful, accommodating and intelligent people inside the space industry. Thanks to you all for sharing your work and your world. The other industry that has supported me is publishing, and I'd like to thank the editors at *Popular Mechanics* (in particular Jim Meigs, David Dunbar, Ryan D'Agostino and Andrew Moseman) for their faith in me over the years. I'd also like to thank Chelsea Cutchens at The Overlook Press for her thoughtful edits and Florida-bred space enthusiasm. No list of gratitude would be complete without saying *grazie* to my father, who still inspires me to be an open-minded and curious traveler, and the rest of my family, who I seldom see but who are more important to me than they know. Above all of these worthy folks, I'd like to thank Alyson Sheppard, the dog-loving, quadcopter-repairing writer/editor who agreed to marry me and make every day special as a rocket launch. This book, like the rest of me, is for her, still and always.

INDEX

B

E

F

G

H

N

R

S

X

Y

Z